AF807585

FM 3-36

Field Manual
No. 3-36

Headquarters
Department of the Army
Washington, DC, 25 February 2009

Electronic Warfare in Operations

Contents

Distribution Restriction: Approved for public release; distribution is unlimited.

Figures

Preface

PURPOSE

FM 3-36 provides Army doctrine for electronic warfare (EW) planning, preparation, execution, and assessment in support of full spectrum operations. Users of FM 3-36 must be familiar with full spectrum operations established in FM 3-0; the military decisionmaking process established in FM 5-0; the operations process established in FMI 5-0.1; commander's visualization described in FM 6-0; and electronic warfare described in JP 3-13.1.

SCOPE

FM 3-36 is organized into seven chapters and six appendixes. Each chapter addresses a major aspect of Army EW operations. The appendixes address aspects of EW operations that complement the operational doctrine. A glossary contains selected terms.

- Chapter 1 discusses the nature and scope of electronic warfare and the impact of the electromagnetic environment on Army operations.
- Chapter 2 offers a discussion of EW support to full spectrum operations, combat power, the warfighting functions, and information tasks.
- Chapter 3 introduces the organizational framework for command and control of EW operations.
- Chapter 4 describes how commanders integrate EW operations throughout the operations process.
- Chapter 5 discusses the coordination required to synchronize and deconflict EW operations effectively.
- Chapter 6 provides the baseline for integrating EW operations into joint and multinational operations.
- Chapter 7 discusses the enabling activities that support EW operations, such as command and control, intelligence, logistics, technical support and EW training.
- Appendix A discusses the electromagnetic environment.
- Appendix B illustrates an EW appendix to an operation order.
- Appendix C illustrates an EW running estimate.
- Appendix D discusses EW related reports and messages.
- Appendix E offers a reference guide to Army and joint EW capabilities.
- Appendix F discusses EW-related tools and resources.

APPLICABILITY

FM 3-36 provides guidance on EW operations for commanders and staffs at all echelons. This FM serves as an authoritative reference for personnel who—

- Develop doctrine (fundamental principles and tactics, techniques, and procedures), materiel, and force structure.
- Develop institutional and unit training.
- Develop standing operating procedures for unit operations.
- Conduct planning, preparation, execution and assessment of electronic warfare.

FM 3-36 applies to the Active Army, Army National Guard/Army National Guard of the United States, and U.S. Army Reserve, unless otherwise stated.

Chapter 1

Electronic Warfare Overview

This chapter provides an overview of electronic warfare and the conceptual foundation that leaders require to understand the electromagnetic environment and its impact on Army operations.

OPERATIONAL ENVIRONMENTS

1-1. An *operational environment* is a composite of the conditions, circumstances, and influences that affect the employment of capabilities and bear on the decisions of the commander (JP 3-0). An operational environment includes physical areas—the air, land, maritime, and space domains. It also includes the information that shapes the operational environment as well as enemy, adversary, friendly, and neutral systems relevant to a joint operation. Joint planners analyze operational environments in terms of six interrelated operational variables: political, military, economic, social, information, and infrastructure. To these variables Army doctrine adds two more: physical environment and time. (See FM 3-0 for additional information on the operational variables). Army leaders use operational variables to understand and analyze the broad environment in which they are conducting operations.

1-2. Army leaders use mission variables to synthesize operational variables and tactical-level information with local knowledge about conditions relevant to their mission. They use mission variables to focus analysis on specific elements that directly affect their mission. Upon receipt of a warning order or mission, Army tactical leaders narrow their focus to six mission variables known as METT-TC. They are mission, enemy, terrain and weather, troops and support available, time available and civil considerations. The mission variables outline the situation as it applies to a specific Army unit.

1-3. Commanders employ and integrate their unit's capabilities and actions within their operational environment to achieve a desired end state. Through analyzing their operational environment, commanders understand how the results of friendly, adversary, and neutral actions may impact that end state. During military operations, both friendly and enemy commanders depend on the flow of information to make informed decisions. This flow of information depends on the electronic systems and devices used to communicate, navigate, sense, store, and process information.

INFORMATION AND THE ELECTROMAGNETIC SPECTRUM

1-4. Commanders plan for and operate electronic systems and the weapon systems that depend on them in an intensive and nonpermissive electromagnetic environment. They ensure the flow of information required for their decisionmaking. (Appendix A further discusses the electromagnetic environment.) Within the electromagnetic environment, electronic systems and devices operate in the electromagnetic spectrum. (See figure 1-1, page 1-2.)

1-5. The electromagnetic spectrum has been used for commercial and military applications for over a century. However, the full potential for its use as the primary enabler of military operations is not yet fully appreciated. New technologies are expanding beyond the traditional radio frequency spectrum. They include high-power microwaves and directed-energy weapons. These new technologies are part of an electronic warfare (EW) revolution by military forces. Just as friendly forces leverage the electromagnetic spectrum to their advantage, so do capable enemies use the electromagnetic spectrum to threaten friendly force operations. The threat is compounded by the growth of a wireless world and the increasingly sophisticated use of commercial off-the-shelf technologies.

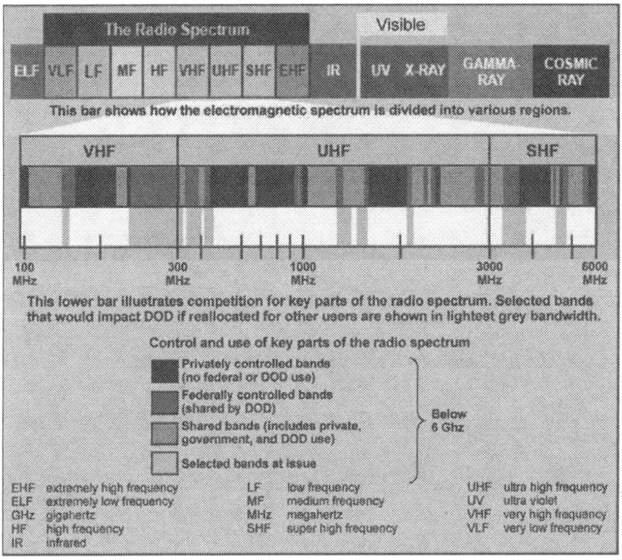

Figure 1-1. The electromagnetic spectrum

1-6. Adversaries and enemies, from small and single actors to large state, multinational, and nonstate actors, use the most modern technology. Such technology is moving into the cellular and satellite communications area. Most military and commercial operations rely on electromagnetic technologies and are susceptible to the inherent vulnerabilities associated with their use. This reliance requires Army forces to dominate the electromagnetic spectrum (within their operational environment) with the same authority that they dominate traditional land warfare operations. Emerging electromagnetic technologies offer expanded EW capabilities. They dynamically affect the electromagnetic spectrum through delivery and integration with other types of emerging weapons and capabilities. Examples are directed-energy weapons, high-powered microwaves, lasers, infrared, and electro-optical and wireless networks and devices.

1-7. In any conflict, commanders attempt to dominate the electromagnetic spectrum. They do this by locating, targeting, exploiting, disrupting, degrading, deceiving, denying, or destroying the enemy's electronic systems that support military operations or deny the spectrum's use by friendly forces. The increasing portability and affordability of sophisticated electronic equipment guarantees that the electromagnetic environment in which forces operate will become even more complex. To ensure unimpeded access to and use of the electromagnetic spectrum, commanders plan, prepare, execute, and assess EW operations against a broad set of targets within the electromagnetic spectrum. (See figure 1-2.)

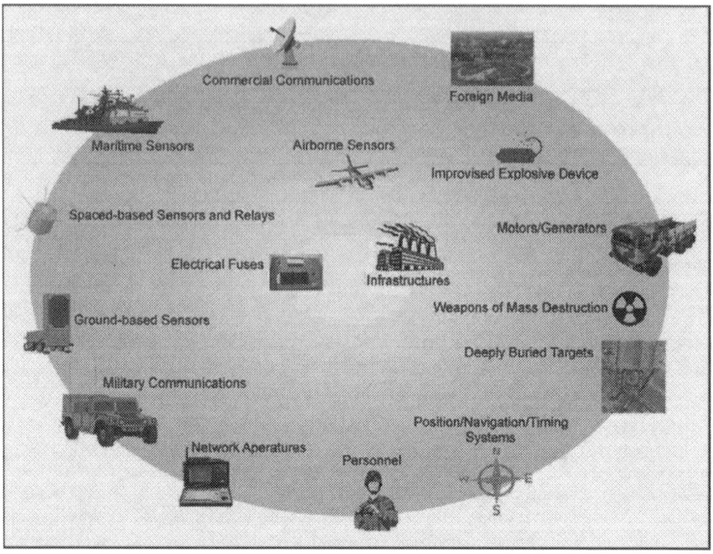

Figure 1-2. Electromagnetic spectrum targets

DIVISIONS OF ELECTRONIC WARFARE

1-8. *Electronic warfare* is defined as military action involving the use of electromagnetic and directed energy to control the electromagnetic spectrum or to attack the enemy. Electronic warfare consists of three divisions: electronic attack, electronic protection, and electronic warfare support (JP 3-13.1). (See figure 1-3.)

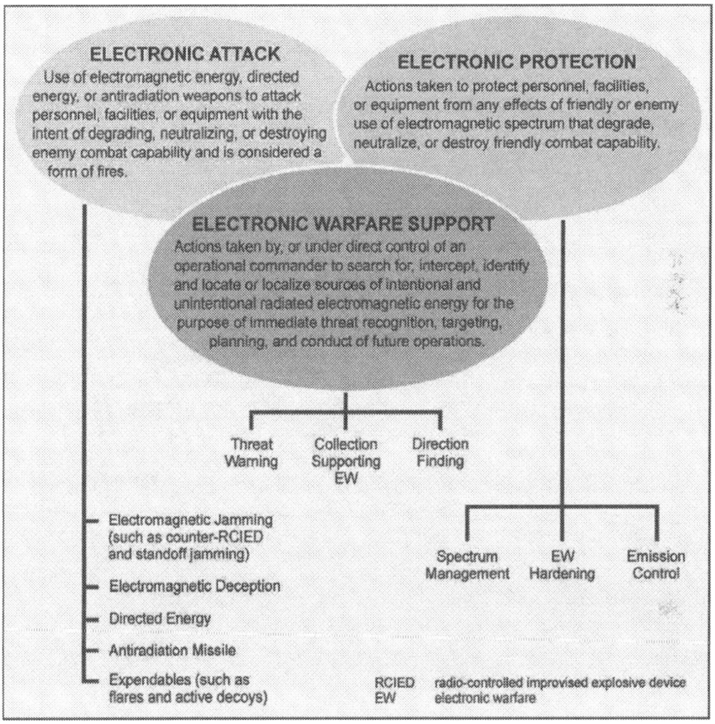

Figure 1-3. The three subdivisions of electronic warfare

ELECTRONIC ATTACK

1-9. *Electronic attack* is a division of electronic warfare involving the use of electromagnetic energy, directed energy, or antiradiation weapons to attack personnel, facilities, or equipment with the intent of degrading, neutralizing, or destroying enemy combat capability and is considered a form of fires (JP 3-13.1). Electronic attack includes—

- Actions taken to prevent or reduce an enemy's effective use of the electromagnetic spectrum, such as jamming and electromagnetic deception.
- Employment of weapons that use either electromagnetic or directed energy as their primary destructive mechanism (lasers, radio frequency weapons, particle beams).
- Offensive and defensive activities including countermeasures.

1-10. Common types of electronic attack include spot, barrage, and sweep electromagnetic jamming. Electronic attack actions also include various electromagnetic deception techniques such as false target or duplicate target generation. (See paragraphs 1-23 to 1-31 for further discussion of electronic attack activities.)

1-11. *Directed energy* is an umbrella term covering technologies that relate to the production of a beam of concentrated electromagnetic energy or atomic or subatomic particles (JP 1-02). A directed-energy weapon uses directed energy primarily as a direct means to damage or destroy an enemy's equipment, facilities, and personnel. In addition to destructive effects, directed-energy weapon systems support area denial and crowd control. (See appendix A for more information on directed energy.)

1-12. Examples of offensive electronic attack include—

- Jamming enemy radar or electronic command and control systems.
- Using antiradiation missiles to suppress enemy air defenses (antiradiation weapons use radiated energy emitted from the target as their mechanism for guidance onto targeted emitters).
- Using electronic deception techniques to confuse enemy intelligence, surveillance, and reconnaissance systems.
- Using directed-energy weapons to disable an enemy's equipment or capability.

1-13. Defensive electronic attack uses the electromagnetic spectrum to protect personnel, facilities, capabilities, and equipment. Examples include self-protection and other protection measures such as use of expendables (flares and active decoys), jammers, towed decoys, directed-energy infrared countermeasure systems, and counter-radio-controlled improvised-explosive-device systems. (See JP 3-13.1 for more discussion of electronic attack.)

ELECTRONIC PROTECTION

1-14. *Electronic protection* is a division of electronic warfare involving actions taken to protect personnel, facilities, and equipment from any effects of friendly or enemy use of the electromagnetic spectrum that degrade, neutralize, or destroy friendly combat capability (JP 3-13.1). For example, electronic protection includes actions taken to ensure friendly use of the electromagnetic spectrum, such as frequency agility in a radio, or variable pulse repetition frequency in radar. Electronic protection should not be confused with self-protection. Both defensive electronic attack and electronic protection protect personnel, facilities, capabilities, and equipment. However, electronic protection protects from the effects of electronic attack (friendly and enemy), while defensive electronic attack primarily protects against lethal attacks by denying enemy use of the electromagnetic spectrum to guide or trigger weapons.

1-15. During operations, electronic protection includes, but is not limited to, the application of training and procedures for countering enemy electronic attack. Army commanders and forces understand the threat and vulnerability of friendly electronic equipment to enemy electronic attack and take appropriate actions to safeguard friendly combat capability from exploitation and attack. Electronic protection measures minimize the enemy's ability to conduct electronic warfare support (electronic warfare support is discussed in paragraphs 1-18 to 1-20) and electronic attack operations successfully against friendly forces. To protect friendly combat capabilities, units—

- Regularly brief force personnel on the EW threat.
- Ensure that electronic system capabilities are safeguarded during exercises, workups, and predeployment training.
- Coordinate and deconflict electromagnetic spectrum usage.
- Provide training during routine home station planning and training activities on appropriate electronic protection active and passive measures.
- Take appropriate actions to minimize the vulnerability of friendly receivers to enemy jamming (such as reduced power, brevity of transmissions, and directional antennas).

1-16. Electronic protection also includes spectrum management. The spectrum manager works for the G-6 or S-6 and plays a key role in the coordination and deconfliction of spectrum resources allocated to the force. Spectrum managers or their direct representatives participate in the planning for EW operations.

1-17. The development and acquisition of communications and electronic systems includes electronic protection requirements to clarify performance parameters. Army forces design their equipment to limit inherent vulnerabilities. If electronic attack vulnerabilities are detected, then units must review these programs. (See DODI 4650.01 for information on the spectrum certification process and electromagnetic compatibility.)

ELECTRONIC WARFARE SUPPORT

1-18. *Electronic warfare support* is a division of electronic warfare involving actions tasked by, or under the direct control of, an operational commander to search for, intercept, identify, and locate or localize sources of intentional and unintentional radiated electromagnetic energy for the purpose of immediate threat recognition, targeting, planning, and conduct of future operations (JP 3-13.1).

1-19. Electronic warfare support systems are a source of information for immediate decisions involving electronic attack, electronic protection, avoidance, targeting, and other tactical employments of forces. Electronic warfare support systems collect data and produce information or intelligence to—

- Corroborate other sources of information or intelligence.
- Conduct or direct electronic attack operations.
- Initiate self-protection measures.
- Task weapon systems.
- Support electronic protection efforts.
- Create or update EW databases.
- Support information tasks.

1-20. Electronic warfare support and signals intelligence missions use the same resources. The two differ in the detected information's intended use, the degree of analytical effort expended, the detail of information provided, and the time lines required. Like tactical signals intelligence, electronic warfare support missions respond to the immediate requirements of a tactical commander. Signals intelligence above the tactical level is under the operational control of the National Security Agency and directly supports the overarching national security mission. Resources that collect tactical-level electronic warfare support data can simultaneously collect national-level signals intelligence. See FM 2-0 for more information on signals intelligence.

ACTIVITIES AND TERMINOLOGY

1-21. Although new equipment and tactics, techniques, and procedures continue to be developed, the physics of electromagnetic energy remains constant. Hence, effective EW activities remain the same despite changes in hardware and tactics. Principal EW activities are discussed in the following paragraphs.

PRINCIPAL ACTIVITIES

1-22. Principal EW activities support full spectrum operations by exploiting the opportunities and vulnerabilities inherent in the use of the electromagnetic spectrum. The numerous EW activities are categorized by the EW subdivisions with which they are most closely associated: electronic attack, electronic warfare support, and electronic protection. JP 3-13.1 discusses these principal activities in detail.

Electronic Attack Activities

1-23. Activities related to electronic attack are either offensive or defensive and include—

- Countermeasures.
- Electromagnetic deception.
- Electromagnetic intrusion.
- Electromagnetic jamming.
- Electromagnetic pulse.
- Electronic probing.

Countermeasures

1-24. *Countermeasures* are that form of military science that, by the employment of devices and/or techniques, has as its objective the impairment of the operational effectiveness of enemy activity (JP 1-02). They can be deployed preemptively or reactively. Devices and techniques used for EW countermeasures include electro-optical-infrared countermeasures and radio frequency countermeasures.

1-25. *Electro-optical-infrared countermeasures* consist of any device or technique employing electro-optical-infrared materials or technology that is intended to impair or counter the effectiveness of enemy activity, particularly with respect to precision guided weapons and sensor systems. Electro-optical-infrared is the part of the electromagnetic spectrum between the high end of the far infrared and the low end of ultraviolet. Electro-optical-infrared countermeasures may use laser and broadband jammers, smokes/aerosols, signature suppressants, decoys, pyrotechnics/pyrophorics, high-energy lasers, or directed infrared energy countermeasures (JP 3-13.1).

1-26. *Radio frequency countermeasures* consist of any device or technique employing radio frequency materials or technology that is intended to impair the effectiveness of or counter enemy activity, particularly with respect to precision guided weapons and sensor systems (JP 3-13.1).

Electromagnetic Deception

1-27. *Electromagnetic deception* is the deliberate radiation, reradiation, alteration, suppression, absorption, denial, enhancement, or reflection of electromagnetic energy in a manner intended to convey misleading information to an enemy or to enemy electromagnetic-dependent weapons, thereby degrading or neutralizing the enemy's combat capability (JP 3-13.4). Among the types of electromagnetic deception are the following:

- Manipulative electromagnetic deception involves actions to eliminate revealing, or convey misleading, electromagnetic telltale indicators that may be used by hostile forces.
- Simulative electromagnetic deception involves actions to simulate friendly, notional, or actual capabilities to mislead hostile forces.
- Imitative electromagnetic deception introduces electromagnetic energy into enemy systems that imitates enemy emissions.

Electromagnetic Intrusion

1-28. *Electromagnetic intrusion* is the intentional insertion of electromagnetic energy into transmission paths in any manner, with the objective of deceiving operators or of causing confusion (JP 1-02).

Electromagnetic Jamming

1-29. *Electromagnetic jamming* is the deliberate radiation, re-radiation, or reflection of electromagnetic energy for the purpose of preventing or reducing an enemy's effective use of the electromagnetic spectrum, with the intent of degrading or neutralizing the enemy's combat capability (JP 1-02).

Electromagnetic Pulse

1-30. *Electromagnetic pulse* is the electromagnetic radiation from a strong electronic pulse, most commonly caused by a nuclear explosion that may couple with electrical or electronic systems to produce damaging current and voltage surges (JP 1-02).

Electronic Probing

1-31. *Electronic probing* is the intentional radiation designed to be introduced into the devices or systems of potential enemies for the purpose of learning the functions and operational capabilities of the devices (JP 1-02). This activity is coordinated through joint or interagency channels and supported by Army forces.

Electronic Warfare Support Activities

1-32. Activities related to electronic warfare support include—

- Electronic reconnaissance.
- Electronic intelligence.
- Electronics security.

Electronic Reconnaisance

1-33. *Electronic reconnaissance* is the detection, location, identification, and evaluation of foreign electromagnetic radiations (JP 1-02).

Electronic Intelligence

1-34. *Electronic intelligence* is technical and geolocation intelligence derived from foreign noncommunications electromagnetic radiations emanating from other than nuclear detonations or radioactive sources (JP 1-02).

Electronics Security

1-35. *Electronics security* is the protection resulting from all measures designed to deny unauthorized persons information of value that might be derived from their interception and study of noncommunications electromagnetic radiations, e.g., radar (JP 1-02).

Electronic Protection Activities

1-36. Activities related to electronic protection include—

- Electromagnetic hardening.
- Electromagnetic interference.
- Electronic masking.
- Electronic warfare reprogramming.
- Emission control.
- Spectrum management.
- Wartime reserve modes.
- Electromagnetic compatibility.

Electromagnetic Hardening

1-37. *Electromagnetic hardening* consists of action taken to protect personnel, facilities, and/or equipment by filtering, attenuating, grounding, bonding, and/or shielding against undesirable effects of electromagnetic energy (JP 1-02).

Electromagnetic Interference

1-38. *Electromagnetic interference* is any electromagnetic disturbance that interrupts, obstructs, or otherwise degrades or limits the effective performance of electronics and electrical equipment. It can be induced intentionally, as in some forms of electronic warfare, or unintentionally, as a result of spurious emissions and responses, intermodulation products and the like (JP 1-02).

Electronic Masking

1-39. *Electronic masking* is the controlled radiation of electromagnetic energy on friendly frequencies in a manner to protect the emissions of friendly communications and electronic systems against enemy electronic warfare support measures/signals intelligence, without significantly degrading the operation of friendly systems (JP 1-02).

Electronic Warfare Reprogramming

1-40. *Electronic warfare reprogramming* is the deliberate alteration or modification of electronic warfare or target sensing systems, or the tactics and procedures that employ them, in response to validated changes in equipment, tactics, or the electromagnetic environment. These changes may be the result of deliberate actions on the part of friendly, adversary, or third parties; or may be brought about by electromagnetic interference or other inadvertent phenomena. The purpose of electronic warfare reprogramming is to maintain or enhance the effectiveness of electronic warfare and target sensing system equipment. Electronic warfare reprogramming includes changes to self-defense systems, offensive weapons systems, and intelligence collection systems (JP 3-13.1).

Emission Control

1-41. *Emission control* is the selective and controlled use of electromagnetic, acoustic, or other emitters to optimize command and control capabilities while minimizing transmissions for operations security: a. detection by enemy sensors; b. mutual interference among friendly systems; and/or c. enemy interference with the ability to execute a military deception plan (JP 1-02).

Electromagnetic Spectrum Management

1-42. *Electromagnetic spectrum management* is planning, coordinating, and managing joint use of the electromagnetic spectrum through operational, engineering, and administrative procedures. The objective of spectrum management is to enable electronic systems to perform their functions in the intended environment without causing or suffering unacceptable interference (JP 6-0).

Wartime Reserve Modes

1-43. *Wartime reserve modes* are characteristics and operating procedures of sensors, communications, navigation aids, threat recognition, weapons, and countermeasures systems that will contribute to military effectiveness if unknown to or misunderstood by opposing commanders before they are used, but could be exploited or neutralized if known in advance. Wartime reserve modes are deliberately held in reserve for wartime or emergency use and seldom, if ever, applied or intercepted prior to such use (JP 1-02).

Electromagnetic Compatibility

1-44. *Electromagnetic compatibility* is the ability of systems, equipment, and devices that utilize the electromagnetic spectrum to operate in their intended operational environments without suffering unacceptable degradation or causing unintentional degradation because of electromagnetic radiation or response. It involves the application of sound electromagnetic spectrum management; system, equipment, and device design configuration that ensures interference-free operation; and clear concepts and doctrines that maximize operational effectiveness (JP 1-02).

Application Terminology

1-45. EW capabilities are applied from the air, land, sea, and space by manned, unmanned, attended, or unattended systems. Units employ EW capabilities to achieve the desired lethal or nonlethal effect on a given target. Units maintain freedom of action in the electromagnetic spectrum while controlling the use of it by the enemy. Regardless of the application, units employing EW capabilities must use appropriate levels of control and protection of the electromagnetic spectrum. In this way, they avoid adversely affecting friendly forces. (Improper EW actions must be avoided because they may cause fratricide or eliminate high-value intelligence targets.)

1-46. In the context of EW application, units use several terms to facilitate control and protection of the electromagnetic spectrum. Terms used in EW application include control, detection, denial, deception, disruption and degradation, protection, and destruction. The three subdivisions of EW—electronic attack, electronic protection, and electronic warfare support—are specified within the following descriptions.

Control

1-47. In the context of EW, control of the electromagnetic spectrum is achieved by effectively coordinating friendly systems while countering enemy systems. Electronic attack limits enemy use of the electromagnetic spectrum. Electronic protection secures use of the electromagnetic spectrum for friendly forces, and electronic warfare support enables the commander's accurate assessment of the situation. All three are integrated for effectiveness. Commanders ensure maximum integration of communications; intelligence, surveillance, and reconnaissance; and information tasks.

Detection

1-48. In the context of EW, detection is the active and passive monitoring of the operational environment for radio frequency, electro-optic, laser, infrared, and ultraviolet electromagnetic threats. Detection is the first step in EW for exploitation, targeting, and defensive planning. Friendly forces maintain the capability to detect and characterize interference as hostile jamming or unintentional electromagnetic interference.

Denial

1-49. In the context of EW, denial is controlling the information an enemy receives via the electromagnetic spectrum and preventing the acquisition of accurate information about friendly forces. Degradation uses traditional jamming techniques, expendable countermeasures, destructive measures, or network applications. These range from limited effects up to complete denial of usage.

Deception

1-50. In the context of EW, deception is confusing or misleading an enemy by using some combination of human-produced, mechanical, or electronic means. Through use of the electromagnetic spectrum, EW deception manipulates the enemy's decision loop, making it difficult to establish accurate situational awareness.

Disruption and Degradation

1-51. In the context of EW, disruption and degradation techniques interfere with the enemy's use of the electromagnetic spectrum to limit enemy combat capabilities. This is achieved with electronic jamming, electronic deception, and electronic intrusion. These enhance attacks on hostile forces and act as force multipliers by increasing enemy uncertainty, while reducing uncertainty for friendly forces. Advanced electronic attack techniques offer the opportunity to nondestructively disrupt or degrade enemy infrastructure.

Protection

1-52. In the context of EW, protection is the use of physical properties; operational tactics, techniques, and procedures; and planning and employment processes to ensure friendly use of the electromagnetic spectrum. This includes ensuring that offensive EW activities do not electronically destroy or degrade friendly intelligence sensors or communications systems. Protection is achieved by component hardening, emission control, and frequency management and deconfliction. Frequency management and deconfliction include the capability to detect, characterize, geolocate, and mitigate electromagnetic interference that affects operations. Protection includes other means to counterattack and defeat enemy attempts to control the electromagnetic spectrum. Additionally, organizations such as a joint force commander's EW staff or a joint EW coordination cell enhance electronic protection by deconflicting EW efforts.

Destruction

1-53. Destruction, in the context of EW, is the elimination of targeted enemy systems. Sensors and command and control nodes are lucrative targets because their destruction strongly influences the enemy's perceptions and ability to coordinate actions. Various weapons and techniques ranging from conventional munitions and directed energy weapons to network attacks can destroy enemy systems that use the electromagnetic spectrum. Electronic warfare support provides target location and related information. While destroying enemy equipment can effectively deny the enemy use of the electromagnetic spectrum, the duration of denial will depend on the enemy's ability to reconstitute. (See JP 3-13.1.)

MEANS VERSUS EFFECTS

1-54. EW means are applied against targets to create a full range of lethal and nonlethal effects. (See figure 1-4.) Choosing a specific EW capability depends on the desired effect on the target and other considerations, such as time sensitivity or limiting collateral damage. EW capabilities provide commanders with additional options for achieving their objectives. During major combat operations there may be circumstances where commanders want to limit the physical damage on a given target. Under such circumstances, the EW staff articulates clearly to the commander the lethal and nonlethal effects EW capabilities can achieve. For example, a target might be enemy radar mounted on a fixed tower. Two EW options to defeat the radar could be to jam the radar or destroy it with antiradiation missiles. If the commander desired to limit damage to the tower, an electronic attack jamming platform would be preferred. In circumstances where commanders cannot sufficiently limit undesired effects such as collateral damage, they may be constrained from applying physical force. The EW staff articulates succinctly how EW capabilities can support actions to achieve desired effects and provide lethal and nonlethal options for commanders.

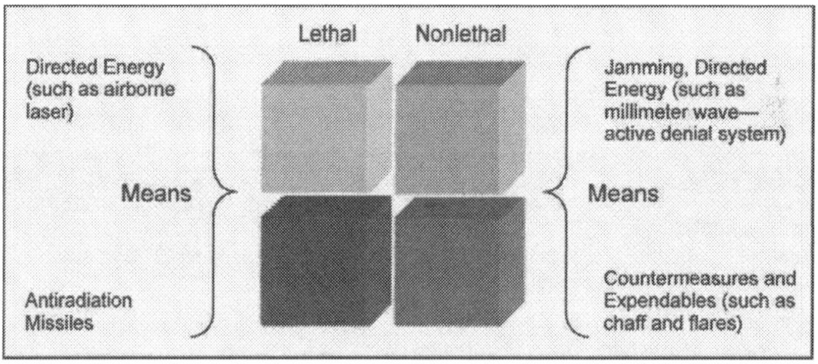

Figure 1-4. Means versus effects

SUMMARY

1-55. As the modern battlefield becomes more technologically sophisticated, military operations continue to be executed in an increasingly complex electromagnetic environment. Therefore, commanders and staffs need to thoroughly understand and articulate how the electromagnetic environment impacts their operations and how friendly EW operations can be used to gain an advantage. Commanders and staffs use the terminology presented in this chapter to describe the application of EW. This ensures a common understanding and consistency within plans, orders, standing operating procedures, and directives.

Chapter 2

Electronic Warfare in Full Spectrum Operations

Information technology is becoming universally available. Most enemies rely on communications and computer networks to make and implement decisions. Radios remain the backbone of tactical military command and control architectures. However, most communications relayed over radio networks are becoming digital as more computers link networks through transmitted frequencies. Therefore, the ability to dominate the electromagnetic spectrum is central to full spectrum operations. This chapter describes how commanders apply electronic warfare capabilities to support full spectrum operations.

THE ROLE OF ELECTRONIC WARFARE

2-1. Army electronic warfare (EW) operations seek to provide the land force commander with capabilities to support full spectrum operations. Full spectrum operations consist of the purposeful, simultaneous combination of offense, defense, and stability or civil support. The goal of full spectrum operations is to change the operational environment so that peaceful processes are dominant. Nonetheless, operational environments are complex; commanders must conduct operations across the entire spectrum of conflict. The Army maintains flexible forces with balanced capabilities and capacities. These flexible and balanced forces remain able to conduct major operations while executing other day-to-day smaller-scale operations. (See FM 3-0.)

2-2. Figure 2-1 (page 2-2) shows the weight of effort for using EW during operations. This figure adapts the elements of full spectrum operations (offense, defense, and stability or civil support) as described in FM 3-0. Overseas, Army forces conduct full spectrum operations (offensive, defensive, and stability) simultaneously as part of a joint force. Within the United States, Army forces conduct homeland defense and civil support operations as part of homeland security. Army electronic warfare (EW) operations seek to provide the land force commander with capabilities to support full spectrum operations. As noted in figure 2-1, statutory law limits the use of EW capabilities in support of civil support operations.

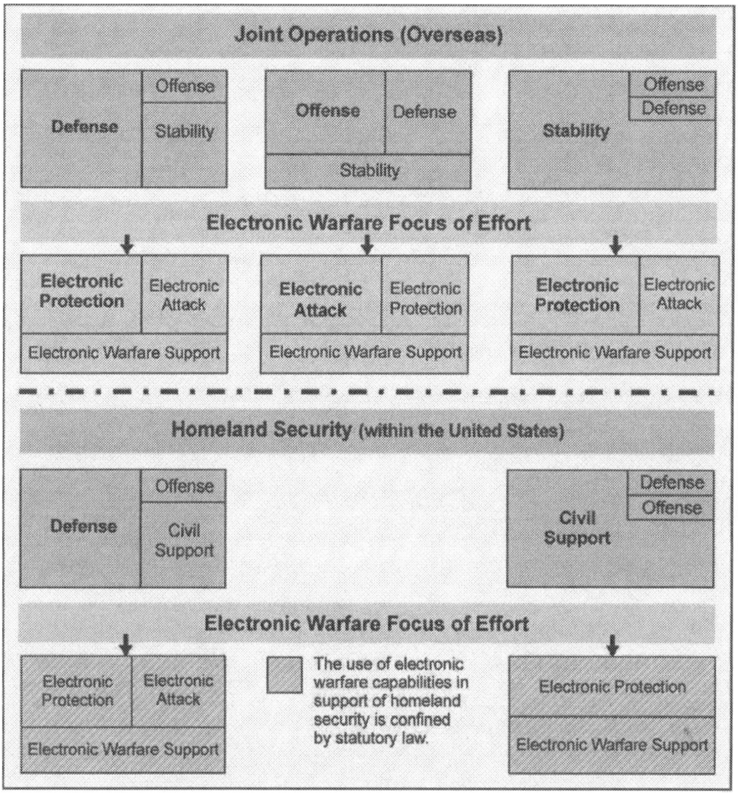

Figure 2-1. Electronic warfare weight of effort during operations

2-3. Full spectrum operations involve more than executing all elements of operations simultaneously. They require that commanders and staffs consider their unit's capabilities and capacities relative to each of the elements of full spectrum operations. Commanders consider how much can be accomplished simultaneously, how much can be phased, and what nonorganic resources may be available to solve problems. The same applies to EW in support of full spectrum operations. Commanders and staffs determine which resident and joint force EW capabilities to leverage in support of each element of full spectrum operations. Weighting the EW focus of effort within each of the elements assists commanders and their staffs in visualizing how EW capabilities can support their operations. Commanders combine offensive, defensive, and stability or civil support operations to seize, retain, and exploit the initiative. As they apply the appropriate level of EW effort to support these elements, commanders can seize, retain, and exploit the initiative within the electromagnetic environment.

THE APPLICATION OF ELECTRONIC WARFARE

2-4. To support full spectrum operations and achieve the goal of electromagnetic spectrum dominance, commanders fully integrate EW capabilities and apply them across the elements of combat power. Leadership and information are applied through, and multiply the effects of, the other six elements of combat power. Paragraphs 2-5 through 2-16 discuss the elements of combat power and how EW capabilities can support them.

IN SUPPORT OF LEADERSHIP

2-5. Leadership initiates the conditions for success. Commanders balance the ability to mass the effects of lethal and nonlethal systems with the requirements to deploy and sustain the units that employ those systems. Generating and maintaining combat power throughout an operation is essential. Today's operational environments require leaders who are competent, confident, and informed in using and protecting combat capabilities that operate within the electromagnetic spectrum. Commanders plan, prepare, execute, and assess EW operations to dominate the electromagnetic spectrum within their operational environment. To accomplish this domination, commanders effectively apply and integrate EW operations across the warfighting functions.

IN SUPPORT OF INFORMATION TASKS AND CAPABILITIES

2-6. Information is the element of combat power consisting of meaningful facts, data, and impressions used to develop a common situational understanding, to enable battle command, and to affect the operational environment. (See FM 3-0 for a discussion of combat power.) In modern conflict, gaining information superiority has become as important as lethal action in determining the outcome of operations. *Information superiority* is the operational advantage derived from the ability to collect, process, and disseminate an uninterrupted flow of information while exploiting or denying an adversary's ability to do the same (JP 3-13). To achieve this operational advantage, Army commanders direct efforts that contribute to information superiority. These efforts fall into four primary areas: Army information tasks; intelligence, surveillance, and reconnaissance; knowledge management; and information management. (See FM 3-0 for a discussion of information superiority.)

2-7. The Army information tasks are used to shape a commander's operational environment. These tasks are information engagement, command and control warfare, information protection, operations security, and military deception. Information capabilities can be used to produce both destructive and constructive effects. For example, destructive actions use information capabilities against the enemy's command and control system and other assets to reduce their combat capability. Constructive actions use information capabilities to inform or influence a particular audience or as a means to affect enemy morale. Although applicable to all elements of full spectrum operations, EW capabilities play a major role in enabling and supporting the execution of the command and control warfare and information protection tasks.

2-8. *Command and control warfare* is the integrated use of physical attack, electronic warfare, and computer network operations, supported by intelligence, to degrade, destroy, and exploit an enemy's or adversary's command and control system or to deny information to it (FM 3-0). It includes operations intended to degrade, destroy, and exploit an enemy's or adversary's ability to use the electromagnetic spectrum and computer and telecommunications networks. *Information protection* is active or passive measures that protect and defend friendly information and information systems to ensure timely, accurate, and relevant friendly information. Information protection denies enemies, adversaries, and others the opportunity to exploit friendly information and information systems for their own purposes (FM 3-0). Table 2-1 shows capabilities, intended effects, staff responsibilities, and functional cells for the command and control warfare and information protection tasks. (For further information on the information tasks, refer to FM 3-0.)

Table 2-1. Two Army information tasks: command and control warfare and information protection

Army Information Tasks	Capabilities	Staff Responsibility	Functional Coordinating Cell	Intended Effects	Integrating Process
Command and Control Warfare	Physical Attack Electronic Attack Computer Network Attack Electronic Warfare Support Computer Network Exploitation	G-3/G-2	Fires	Degrade, disrupt, destroy and exploit enemy command and control	Operations Process
Information Protection	Information Assurance Computer Network Defense Electronic Protection	G-6	Network Operations	Protect friendly computer networks and communication means	
G-2 assistant chief of staff, intelligence G-3 assistant chief of staff, operations				G-6 assistant chief of staff, signal	

2-9. To support these information tasks, commanders ensure EW is coordinated, integrated, and synchronized with all other tasks. This occurs within the operations process through the various functional and integrating cells. Table 2-2 illustrates EW capabilities, actions, and objectives that support the command and control warfare and information protection tasks.

Table 2-2. Electronic warfare support to two Army information tasks

Information Tasks	Command and Control Warfare	Information Protection
Electronic warfare supports by	Locating and identifying threat command and control systems. Denying, disrupting, degrading, and/or destroying the enemy's command and control system. Supporting and complementing computer network attack and computer network exploitation operations.	Deconflicting spectrum usage with the spectrum manager. Hardening equipment against electromagnetic interference. Emissions control.
Action	Electronic attack (jamming, antiradiation missiles). Directed energy and electromagnetic spectrum area denial systems. Expendables (chaff, decoys, and flares). Electronic warfare support/signals intelligence.	Frequency agility in radios. Electronic shielding for systems. Electronic masking. Processes to counter intrusion. Implementing emissions control procedures to safeguard friendly systems and facilities from the effects of friendly and enemy electronic attack.
Objective or Effect	Detect, deny, disrupt or degrade, and destroy.	Control and protection.

IN SUPPORT OF THE WARFIGHTING FUNCTIONS

2-10. EW capabilities support each of the six warfighting functions. Examples of specific supporting capabilities are given in the following paragraphs.

Movement and Maneuver

2-11. The *movement and maneuver warfighting function* is the related tasks and systems that move forces to achieve a position of advantage in relation to the enemy. Direct fire is inherent in maneuver, as is close combat (FM 3-0). EW capabilities that enable the movement and maneuver of Army forces include—

- Suppression and destruction of enemy integrated air defenses.
- Denial of enemy information systems and intelligence, surveillance, and reconnaissance sensors.
- Target designation and range finding.
- Protection from effects of friendly and enemy EW.
- Lethal and nonlethal effects against enemy combat capability (personnel, facilities, and equipment).
- Threat warning and direction finding.
- Use of the electromagnetic spectrum to counter improvised explosive device operations.
- Electromagnetic spectrum obscuration, low observability, and multispectral stealth.

Intelligence

2-12. The *intelligence warfighting function* is the related tasks and systems that facilitate understanding of the operational environment, enemy, terrain, and civil considerations (FM 3-0). It includes tasks associated with intelligence, surveillance, and reconnaissance. EW capabilities that enable the intelligence warfighting function include—

- Increased access for intelligence collection assets (systems and personnel) by reducing antiaccess, antipersonnel, and antisystems threats.
- Increased capability to search for, intercept, identify, and locate sources of radiated electromagnetic energy in support of targeting, information tasks, and future operations.
- Increased capability in providing threat recognition and threat warning to the force.
- Indications and warning of threat emitters and radar.
- Denial and destruction of counter-intelligence, -surveillance, and -reconnaissance systems.

Fires

2-13. The *fires warfighting function* is the related tasks and systems that provide collective and coordinated use of Army indirect fires, joint fires, and command and control warfare, including nonlethal fires, through the targeting process (FM 3-0). It includes tasks associated with integrating command and control warfare. EW capabilities that enable the fires warfighting function include—

- Detection and location of targets radiating electromagnetic energy.
- Disruption, degradation, and destruction options for servicing targets. This includes information systems, targets requiring precision strike (such as minimal collateral damage and minimal weapons signature), hard and deeply buried targets, weapons of mass destruction, and power generation and infrastructure targets.
- Control, dispersion, or neutralization of combatant and noncombatant personnel with nonpersistent effects and minimum collateral damage (scalable and nonlethal).
- Area denial capabilities against vehicles, vessels, and aircraft.

Sustainment

2-14. The *sustainment warfighting function* is the related tasks and systems that provide support and services to ensure freedom of action, extend operational reach, and prolong endurance (FM 3-0). EW capabilities that enable the sustainment warfighting function include—

- Protection of sustainment forces from friendly and adversary use of EW in static or mobile environments.
- Enhanced electromagnetic environment situational awareness through the interception, detection, identification, and location of adversary electromagnetic emissions and by providing indications and warnings. (This information can assist in convoy planning, asset tracking, and targeting of potential threats to sustainment operations.)
- Countering improvised explosive devices to support ground lines of communication (includes counter-radio-controlled improvised-explosive-device systems and countering other threats triggered through the electromagnetic spectrum, such as lasers).
- Spectrum deconfliction and emissions control procedures in support of sustainment command and control.
- Electromagnetic spectrum obscuration, low-observability, and multispectral stealth (These capabilities provide protection during sustainment operations).

Command and Control

2-15. The *command and control warfighting function* is the related tasks and systems that support commanders in exercising authority and direction (FM 3-0). EW capabilities that enable the command and control warfighting function include—

- Protection of friendly critical information systems and command and control nodes, personnel, and facilities from the effects of friendly and adversary EW operations.
- Control of friendly EW systems through—
  - Frequency deconfliction.
  - Asset tracking.
  - Employment execution.
  - Reprogramming of EW systems.
  - Registration of all electromagnetic spectrum emitting devices with the spectrum manager (both prior to deployment and when new systems or devices are added to the deployed force).
- The development of EW command and control tools to enhance required coordination between Army and joint EW operations.
- EW operations integration, coordination, deconfliction, and synchronization through the EW working group (see chapter 3).
- Increased commander situational understanding through improved common operational picture input of electromagnetic spectrum- and EW-related information.
- EW operations monitoring and assessment.

Protection

2-16. The *protection warfighting function* is the related tasks and systems that preserve the force so the commander can apply maximum combat power (FM 3-0). EW capabilities and actions that enable the protection warfighting function include—

- Enhanced electromagnetic spectrum situational awareness through the interception, detection, identification, and location of adversary electromagnetic emissions used to providing indications and warnings of threat emitters and radars.
- Denial, disruption, or destruction of electromagnetic-spectrum-triggered improvised explosive devices and enemy air defense systems.
- Deception of enemy forces.
- Electromagnetic spectrum obscuration, low-observability, and multispectral stealth.
- EW countermeasures for platform survivability (air and ground).
- Area denial capabilities (lethal and nonlethal) against personnel, vehicles, and aircraft.
- Protection of friendly personnel, equipment, and facilities from friendly and enemy electronic attack, including friendly information systems and information. (This includes the coordination and use of both airborne and ground-based electronic attack with higher and adjacent units.)

SUMMARY

2-17. Army EW operations provide the land force commander capabilities to support full spectrum operations (offensive, defensive, and stability or civil support operations). EW supports full spectrum operations by applying EW capabilities to detect, deny, deceive, disrupt, or degrade and destroy enemy combat capability and by controlling and protecting friendly use of the electromagnetic spectrum. These capabilities—when applied across the warfighting functions—enable commanders to address a broad set of electromagnetic-spectrum-related targets to gain and maintain an advantage within the electromagnetic spectrum.

Chapter 3

Electronic Warfare Organization

A flexible organizational framework and capable, proficient electronic warfare personnel enable the commander's electronic warfare capability on the battlefield. This chapter discusses a framework that ensures coordination, synchronization, and integration of electronic warfare into full spectrum operations. This electronic warfare organizational framework supports current operations and is adaptable for future operations.

ORGANIZING ELECTRONIC WARFARE OPERATIONS

3-1. Operational challenges across the electromagnetic spectrum are expanding rapidly. As Army electronic warfare (EW) capabilities expand to meet these challenges, the organizational design required to coordinate, synchronize, integrate, and deconflict these capabilities must transform as rapidly. To meet current and future requirements, command and control of EW operations is built around the concept of EW working groups. Figure 3-1, page 3-2, illustrates the EW coordination organizational framework.

ARMY SERVICE COMPONENT COMMAND, CORPS, AND DIVISION LEVELS

3-2. A *working group* is a temporary grouping of predetermined staff representatives who meet to coordinate and provide recommendations for a particular purpose or function (FMI 5-0.1). The EW working group, when established, is responsible to the G-3 through the fires cell. An EW working group usually includes representation from the G-2, G-3, G-5, G-6, and G-7. (Joint doctrine calls this organization the EW coordination cell.) The EW working groups depicted in figure 3-1 (page 3-2) facilitate the internal (Army) and external (joint) integration, synchronization, and deconfliction of EW actions with fires, command and control, movement and maneuver, intelligence, sustainment and protection warfighting functions. Normally, EW working groups do not add additional structure to an existing organization. As depicted in figure 3-1, working groups vary in size and composition based on echelon.

3-3. Normally, the senior EW officer heads the EW working group and is accountable to the G-3 for integrating EW requirements. Working within the fires cell, the EW officer coordinates directly with the fire support coordinator for the integration of EW into the targeting process. This ensures EW capabilities are fully integrated with all other effects. Additional staff representation within EW working groups may include a fire support coordinator, a spectrum manager, a space operations officer, and liaison officers as required. Depending on the echelon, liaisons could include joint, interagency, and multinational representatives. When an Army headquarters serves as the headquarters of a joint task force or joint force land component command, the Army headquarters' working group becomes the joint force EW coordination cell.

3-4. When Army forces are employed as part of a joint or multinational force, they normally have EW representatives supporting higher headquarters' EW coordination organizations. These organizations may include the joint force commander's EW staff or the information operations cell within a joint task force. Sometimes a component EW organization may be designated as the joint EW coordination cell. (Chapter 6 discusses joint electronic warfare operations in more detail.) The overall structure of the combatant force and the level of EW to be conducted determine the structure of the joint EW coordination cell. The organization to accomplish the required EW coordination and functions varies by echelon.

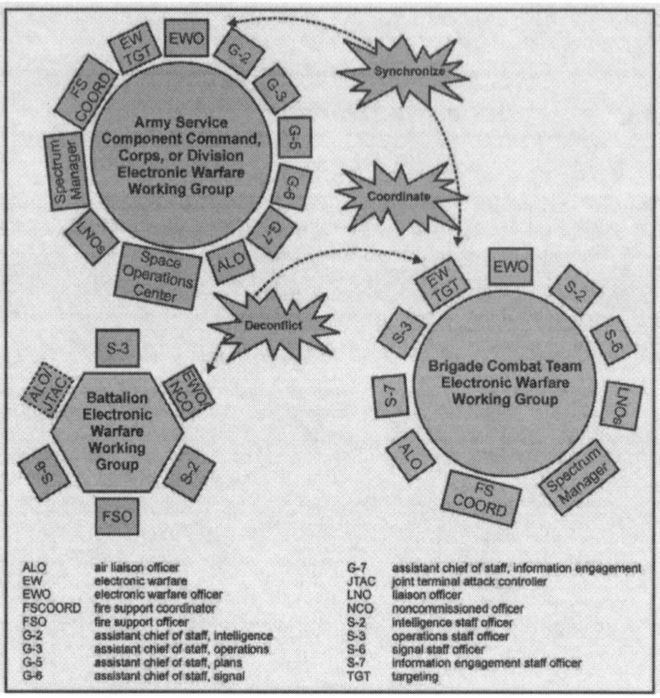

Figure 3-1. Electronic warfare coordination organizational framework

3-5. Regardless of the organizational framework employed, EW working groups perform specific tasks. Table 3-1 (page 3-3) details the functions of the EW working groups by echelon from battalion to Army Service component command. There is no formal organizational framework for EW at the company level (see paragraph 3-9).

Table 3-1. Functions of electronic warfare working groups

EW Working Group	Functions
Division and Above ALO EWO EW targeting G-2 G-3 G-5 G-6 G-7 FSCOORD LNOs Spectrum manager Space support officer	**Peacetime: Division—ASCC** • Conduct long-range electronic warfare planning in support of theater or combatant command requirements. • Integrate electronic warfare into operation plans and concept plans. • Develop electronic warfare supporting plans to operation plans and contingency plans. • Coordinate joint electronic warfare training and exercises. • Develop information and knowledge necessary to support contingency planning (for example, joint restricted frequency list development, spectrum management, and deconfliction). **Wartime: Division—ASCC** • Serve as the joint force land component or joint task force EW working group. • When directed, serve as the jamming control authority. • Develop and promulgate electronic warfare policies and support higher level policies. • Identify and coordinate intelligence support requirements for electronic warfare. • Plan, coordinate, and assess offensive and defensive electronic warfare requirements. • Plan, coordinate, synchronize, deconflict, and assess electronic warfare operations. • Maintain current assessment of electronic warfare resources available to the commander. • Prioritize electronic warfare effects and targets. • Predict effects of friendly and enemy electronic warfare. • Coordinate spectrum management and radio frequency deconfliction with G-6 and J-6. • Plan, assess, and implement friendly electronic security measures. • Plan, coordinate, integrate, and deconflict electronic warfare effects within the operations process.
Brigade S-3 EWO EW targeting FSCOORD S-2 S-6 ALO LNOs S-7 Spectrum manager	**Peacetime:** • Develop electronic warfare supporting requirements to operations plans and exercises. **Wartime:** • Support electronic warfare policies. • Plan, prepare, execute, and assess electronic warfare operations. • Integrate electronic warfare intelligence preparation of the battlefield into the operations process. • Identify and coordinate intelligence support requirements for BCT and subordinate units' electronic warfare operations. • Assess offensive and defensive electronic warfare requirements. • Maintain current assessment of electronic warfare resources available to unit. • Prioritize BCT and subordinate units' electronic warfare targets. • Plan, coordinate, and assess friendly electronic warfare operations. • Implement friendly electronic security measures (for example, electromagnetic spectrum mitigation and network protection). • When directed, serve as the jamming control authority.
Battalion EWO/NCO FSO S-2 S-3 S-6 JTAC	**Peacetime:** • Support BCT electronic warfare requirements to operations and exercises. **Wartime:** • Evaluate electronic warfare offensive, defensive, and support requirements. • Coordinate electronic warfare operations with higher headquarters. • Identify and coordinate intelligence support requirements with higher headquarters. • Execute electronic warfare in support of current operations. • Assess electronic warfare operations.

ALO	air liaison officer	G-7	assistant chief of staff, information engagement
ASCC	Army service component command	J-6	communications system directorate of a joint staff
BCT	brigade combat team		
EW	electronic warfare	JTF	joint task force
EWO	electronic warfare officer	JTAC	joint terminal attack controller
FSCOORD	fire support coordinator	LNO	liaison officer
G-2	assistant chief of staff, intelligence	NCO	noncommissioned officer
G-3	assistant chief of staff, operations	S-2	intelligence staff officer
G-4	assistant chief of staff, logistics	S-3	operations staff officer
G-5	assistant chief of staff, plans	S-6	signal staff officer
G-6	assistant chief of staff, signal	S-7	information engagement staff officer

Brigade Level

3-6. At the brigade level, the EW officer heads the EW working group and is accountable to the S-3 for integrating EW requirements. Additional staff representation within EW working groups at the brigade combat team level may include the fire support coordinator, EW targeting technician, S-2, S-6, spectrum manager, S-7, and liaison officers as required.

3-7. The EW working group at the brigade combat team coordinates with the higher echelon EW working groups. The brigade working group plays an important role in requesting and integrating joint air and ground EW support. It also manages the brigade's organic EW "fight" within the fires cell. The EW officer works as part of the brigade combat team staff. In this position, the EW officer synchronizes, integrates, and deconflicts brigade combat team EW actions with the EW working group at division level. Although EW falls under the control of the S-3, EW officers are fully immersed in fires targeting and planning to ensure proper use and coordination of EW. See table 3-1, page 3-3, for an outline of the functions of the brigade combat team EW working group.

Battalion Level

3-8. At the battalion level, the EW officer or noncommissioned officer leads the EW working group and is accountable to the S-3 for integrating EW requirements. Additional staff representation within EW working groups at the battalion level may include the S-2, S-6, fire support officer, and a joint terminal attack controller when assigned. The battalion EW working group coordinates battalion EW operations with the brigade combat team EW working group. See table 3-1, page 3-3, for an outline of the functions of the battalion EW working group.

Company Level

3-9. At the company level, trained EW personnel holding an additional skill identifier of 1K (tactical EW operations) or 1J (operational EW operations) perform several tasks. They advise the commander on the employment of EW equipment, track EW equipment status, assist operators in the use and maintenance of EW equipment, and coordinate with higher headquarters EW working groups.

PLANNING AND COORDINATING ELECTRONIC WARFARE ACTIVITIES

3-10. Key personnel involved in the planning and coordination of EW activities are—

- G-3 and S-3 staff.
- EW officer.
- Fire support coordinator.
- G-2 and S-2 staff.
- G-6 and S-6 staff.
- Electromagnetic spectrum manager.
- Liaisons.

G-3 or S-3 Staff

3-11. The G-3 or S-3 staff is responsible for the overall planning, coordination, and supervision of EW activities, except for intelligence. The EW officer is part of the G-3 or S-3 staff. The G-3 or S-3 staff—

- Plans for and incorporates EW into operation plans and orders, in particular within the fire support plan and the information operations plan (in joint operations).
- Tasks EW actions to assigned and attached units.
- Exercises control over electronic attack, including integration of electromagnetic deception plans.

- Directs electronic protection measures the unit will take based on recommendations from the G-6 or S-6, the EW officer, and the EW working group.
- Coordinates and synchronizes EW training with other unit training requirements.
- Coordinates and synchronizes EW training with other unit training requirements.
- Issues EW support tasks within the unit intelligence, surveillance, and reconnaissance plan. These tasks are according to the collection plan and the intelligence synchronization matrices developed by the G-2 or S-2 and the collection manager.
- Coordinates with the EW working group to ensure planned EW operations support the overall tactical plan.
- Integrates electronic attack as a form of fires within the fires cell.

ELECTRONIC WARFARE OFFICER

3-12. As a member of the G-3 or S-3 staff, the EW officer plans, coordinates, and supports the execution of EW. The EW officer—

- Leads the EW working group.
- Plans, coordinates, and assesses EW offensive, defensive, and support requirements.
- Supports the G-2 or S-2 during intelligence preparation of the battlefield.
- Supports the fire support coordinator to ensure electronic attack fires are integrated with all other effects.
- Plans, assesses, and implements friendly electronics security measures.
- Prioritizes EW effects and targets with the fire support coordinator.
- Plans and coordinates EW operations across functional and integrating cells.
- Deconflicts EW operations with the spectrum manager.
- Maintains a current assessment of available EW resources.
- Participates in other cells and working groups (as required) to ensure EW integration.
- Serves as EW subject matter expert on existing EW rules of engagement.
- When designated, serves as the jamming control authority.
- Prepares, submits for approval, and supervises the issuing and implementation of fragmentary orders for EW operations.

G-2 OR S-2 STAFF

3-13. The G-2 or S-2 staff advises the commander and staff on the intelligence aspects of EW. The G-2 or S-2 staff—

- Provides threat data to support programming of unit EW systems and deconfliction of their use by the EW working group.
- Ensures that electronic order of battle requirements are included in the intelligence collection plan.
- Determines enemy EW organizations, disposition, capabilities, and intentions via collection and analysis.
- Determines enemy EW vulnerabilities and high-value targets.
- Assesses effects of friendly EW operations on the enemy.
- Helps prepare the intelligence-related portion of the EW running estimate.
- Provides input to the restricted frequency list by recommending guarded frequencies.
- Provides updates on the rapid electronic order of battle.
- Maintains appropriate threat EW databases.
- Works with the EW working group to ensure that intelligence collection is synchronized with EW requirements and deconflicted with planned EW actions. Ensures that EW threat data is deconflicted with friendly electromagnetic spectrum needs.

Network Operations Officer

3-14. The network operations officer (in the G-6 or S-6 staff) coordinates the communications network for the following services:

- Preparing the electronic protection policy on behalf of the commander.
- Assisting in preparing EW plans and orders.
- Reporting all enemy electronic attack activity detected by friendly communications and electronics elements to the EW working group for counteraction.
- Assisting the unit EW officer with resolving EW systems maintenance and communications fratricide problems.

Spectrum Manager

3-15. The spectrum manager coordinates electromagnetic spectrum use for a wide variety of communications and electronic resources. The spectrum manager—

- Issues the signal operating instructions.
- Provides all spectrum resources to the task force.
- Coordinates for spectrum usage with higher echelon G-6 or S-6, and applicable host-nation and international agencies as necessary.
- Coordinates the preparation of the restricted frequency list and issuance of emissions control guidance.
- Coordinates frequency allotment, assignment, and use.
- Coordinates electromagnetic deception plans and operations in which assigned communications resources participate.
- Coordinates measures to reduce electromagnetic interference.
- Coordinates with higher echelon spectrum managers for electromagnetic interference resolution that cannot be resolved internally.
- Assists the EW officer in issuing guidance in the unit (including subordinate elements) regarding deconfliction and resolution of interference problems between EW systems and other friendly systems.
- Participates in the EW working group to deconflict friendly electromagnetic spectrum requirements with planned EW operations and intelligence collection.

SUMMARY

3-16. The organizational framework for EW coordination and functions varies by echelon. The necessity to form an EW working group is largely based on the overall structure of the combatant force and the level of EW to be conducted. During unified actions, other Service EW officers, signals intelligence officers, and EW asset representatives are invaluable to Army EW working groups in the planning, preparation, execution, and assessment of EW operations. As Army EW capabilities and concepts for employment continue to evolve, so do the organizational designs that ensure their effective command and control and execution in support of operations.

Chapter 4

Electronic Warfare and the Operations Process

The *operations process* consists of the major command and control activities performed during operations: planning, preparing, executing, and continuously assessing the operation. The commander drives the operations process (FM 3-0). These activities occur continuously throughout an operation, overlapping and recurring as required (see figure 4-1). The staff electronic warfare officer is actively involved in the operations process. Electronic warfare planning, preparation, execution, and assessment require collective expertise from operations, intelligence, signal, and battle command. The electronic warfare officer—through the unit's electronic warfare working group—integrates efforts across the warfighting functions. This ensures that electronic warfare operations support the commander's objectives.

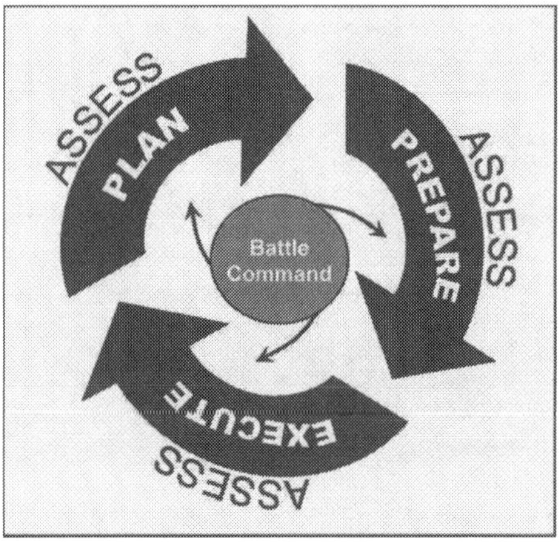

Figure 4-1. The operations process

SECTION I — ELECTRONIC WARFARE PLANNING

4-1. Electronic warfare (EW) planning is based on three main considerations. The first is applying the military decisionmaking process (MDMP). EW planners understand and follow its seven steps. In a time-constrained environment they still follow all seven steps, abbreviating the MDMP process appropriately. Additionally, EW planners apply EW integrating processes. They understand how EW actions contribute to operations. They integrate and synchronize EW activities starting with planning and continuing throughout operations. Finally, EW planners apply EW employment considerations according to the characteristics of EW capabilities.

THE MILITARY DECISIONMAKING PROCESS

4-2. EW planning minimizes fratricide and optimizes operational effectiveness during execution. Therefore, EW planning occurs concurrently with other operational planning during the MDMP. The MDMP synchronizes several processes, including intelligence preparation of the battlefield (IBP) (see FM 34-130), the targeting process (see FM 6-20-10), and risk management (see FM 5-19). These processes occur continuously during operations.

4-3. Depending on the organizational echelon, the staff EW officer leads EW planning through the EW working group. (The EW working group at echelons above brigade is sometimes referred to as an EW coordination cell.) An EW working group is normally supported by representatives from the G-2 or S-2, G-3 or S-3, G-6 or S-6, and other staff as required. Other staff representatives can include the fire support coordinator or fire support officer, spectrum manager, air liaison officer, space officer, and liaison officers. Paragraphs 4-5 through 4-33 outline key EW contributions to the processes and planning actions that occur during the seven steps of the MDMP. (FM 5-0 discusses the MDMP.)

RECEIPT OF MISSION

4-4. Commanders begin the MDMP upon receiving or anticipating a new mission. During this first step, commanders issue their initial guidance and initial information requirements or commander's critical information requirements.

4-5. Upon receipt of a mission, the staff EW officer alerts the staff members supporting the EW working group. The EW officer and support staff begin to gather the resources required for mission analysis. Resources might include a higher headquarters operation order or plan, maps of the area of operations, electronic databases, required field manuals and standing operating procedures, current running estimates, and reachback resources (see appendix F). The EW officer also provides input to the staff's initial assessment and updates the EW running estimate. As part of this update, the EW officer identifies all friendly EW assets and resources and their status. The EW officer also provides this information throughout the operations process. This includes monitoring, tracking, and seeking out information relating to EW operations to assist the commander and staff.

MISSION ANALYSIS

4-6. Planning includes a thorough mission analysis. Both the process and products of mission analysis help commanders refine their situational understanding and determine their restated mission. (See FM 5-0 for more details.) The EW officer and supporting members of the EW working group contribute to the overall mission analysis by participating in IPB and through the planning actions discussed in paragraphs 4-7 through 4-14. (Paragraphs 4-35 to 4-40 discuss EW input to IPB during operations.)

4-7. The EW officer and EW working group members—

- Convene the appropriate EW working group.
- Determine known facts, status, or conditions of forces capable of EW operations as defined in the commander's planning documents, such as a warning order or operation order.
- Identify EW planning support requirements and develop support requests as needed.

4-8. The EW officer and EW working group members support the G-2 and S-2 in IPB by—

- Determining the threat's dependence on the electromagnetic spectrum.
- Determining the threat's EW capability.

- Determining the threat's intelligence system collection capability.
- Determining which threat vulnerabilities relate to the electromagnetic spectrum.
- Determining how the operational environment affects EW operations using the operational variables and mission variables as appropriate.
- Initiating, refining, and validating information requirements and requests for information.

4-9. The EW officer and EW working group members—

- Determine facts and develop necessary assumptions relevant to EW such as the status of EW capability at probable execution and time available.
- Analyze the commander's mission and intent from an EW perspective.
- Identify constraints relevant to EW—
  - Actions EW operations must perform.
  - Actions EW operations cannot perform.
  - Other constraints.
- Analyze mission, enemy, terrain and weather, troops and support available, time available and civil considerations from the EW perspective.

4-10. The EW officer and EW working group members determine enemy and friendly centers of gravity and list their critical capabilities, requirements, and vulnerabilities from an EW perspective. (They determine how EW capabilities can best attack an enemy's command and control system.) The center of gravity analysis process outlined in figure 4-2 helps identify and list the critical vulnerabilities of enemy centers of gravity. The EW officer and EW working group members also list the critical requirements associated with the identified command and control critical capability (or command and control nodes) and then identify the critical vulnerabilities associated with the critical requirements. Through this process, the EW officer and EW working group members help determine which vulnerabilities can be engaged by EW capabilities to produce a decisive outcome.

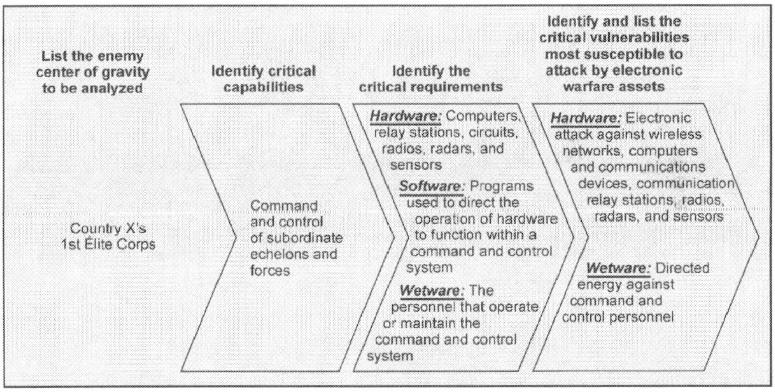

Figure 4-2. Example of analysis for an enemy center of gravity

4-11. Additionally, the EW officer and EW working group members determine how EW can help protect friendly centers of gravity. The center of gravity analysis process as outlined in figure 4-2 can also be used help identify critical vulnerabilities of friendly centers of gravity. The EW officer and EW working group members list the critical requirements associated with the identified friendly command and control critical capability. Then, the EW officer and EW working group members identify the critical vulnerabilities associated with the critical requirements. These vulnerabilities can help determine how to best use EW capabilities to defend or protect friendly centers of gravity from enemy attack. Key to this portion of the analysis is to assess the potential impact of EW operations on friendly information systems such as electromagnetic interference.

4-12. The EW officer and EW working group members identify and list—

- High-value targets that can be engaged by EW capabilities.
- Tasks that EW forces perform according to EW subdivision (electronic attack, electronic warfare support, and electronic protection) in support of the warfighting functions. These include—
  - Determining specified EW tasks.
  - Determining implied EW tasks.

4-13. The EW officer and EW working group members—

- Conduct initial EW force structure analysis to determine if sufficient assets are available to perform the identified EW tasks. (If organic assets are insufficient, they draft requests for support and augmentation.)
- Conduct an initial EW risk assessment and review the risk assessment done by the entire working group.
- Provide EW perspective in the development of the commander's restated mission.
- Assist in development of the mission analysis briefing for the commander.

4-14. By the conclusion of mission analysis, the EW officer and EW working group members generate or gather the following products and information:

- The initial information requirements for EW operations.
- A rudimentary command and control nodal analysis of the enemy.
- The list of EW tasks required to support the mission.
- A list of assumptions and constraints related to EW operations.
- The planning guidance for EW operations.
- EW personnel augmentation or support requirements.
- An update of the EW running estimate.
- EW portion or input to the commander's restated mission.

Course of Action Development

4-15. After receiving the restated mission, commander's intent, and commander's planning guidance, the staff develops courses of action (COAs) for the commander's approval. Figure 4-3 depicts the required input to COA development and identifies the key contributions made by the EW officer and EW working group members during the process and output stages (center and right of figure 4-3). The actions the EW officer and EW working group members perform to support COA development are discussed in more detail in paragraphs 4-16 through 4-20.

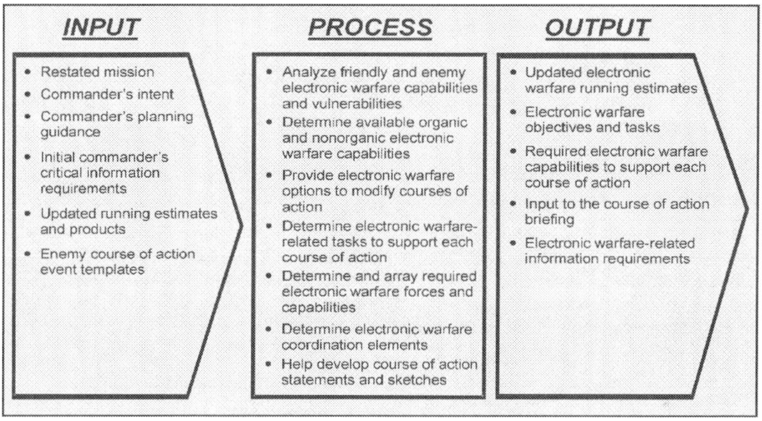

Figure 4-3. Course of action development

4-16. The EW officer and EW working group members contribute to COA development through the following planning actions—

- Determining which friendly EW capabilities are available to support the operation, including organic and nonorganic capabilities for planning.
- Determining possible friendly and enemy EW operations, including identifying friendly and enemy vulnerabilities.

4-17. Additionally, the EW officer and EW working group members help develop initial COA options by—

- Identifying COA options that may be feasible based on their functional expertise (while brainstorming of COAs).
- Providing options to modify a COA to enable accomplishing a requirement within the EW area of expertise.
- Identifying information (relating to EW options) that may impact other functional areas and sharing that information immediately.
- Identifying the EW-related tasks required to support the COA options.

4-18. The EW officer and EW working group members determine the forces required for mission accomplishment by—

- Determining the EW tasks that support each COA and how to perform those tasks based on available forces and capabilities. (Available special technical operations capabilities are considered in this analysis.)
- Providing input and support to proposed deception options.
- Ensuring the EW options provided in support of all possible COAs meet the established screening criteria.

4-19. The EW officer and EW working group members identify EW supporting tasks and their purpose in supporting any decisive, shaping, and sustaining operations as each COA is developed. These EW tasks include those—

- Focused on defeating the enemy.
- Required to protect friendly force operations.

4-20. The EW officer and EW working group members assist in developing the COA briefing as required. By the conclusion of COA development, the EW officer and EW working group members generate or gather the following products and information:

- A list of EW objectives and desired effects related to the EW tasks.
- A list of EW capabilities required to perform the stated EW tasks for each COA.
- The information and intelligence requirements for performing the EW tasks in support of each COA.
- An update to the EW running estimate.

COURSE OF ACTION ANALYSIS (WAR-GAMING)

4-21. The COA analysis allows the staff to synchronize the elements of combat power for each COA and to identify the COA that best accomplishes the mission. It helps the commander and staff to—

- Determine how to maximize the effects of combat power while protecting friendly forces and minimizing collateral damage.
- Further develop a visualization of the battle.
- Anticipate battlefield events.
- Determine conditions and resources required for success.
- Determine when and where to apply force capabilities.
- Focus IPB on enemy strengths and weaknesses as well as the desired end state.
- Identify coordination needed to produce synchronized results.
- Determine the most flexible COA.

Paragraphs 4-22 to 4-23 discuss actions the EW officer and EW working group members perform to support COA analysis. (See FM 5-0 for more information on war-gaming.)

4-22. During COA analysis, the EW officer and EW working group members synchronize EW actions and assist the staff in integrating EW capabilities into each COA. The EW officer and EW working group members address how each EW capability supports each COA. They apply these capabilities to associated time lines, critical events, and decision points in the synchronization matrix (see table 4-1). During this planning phase, the EW officer and EW working group members aim to—

- Analyze each COA from an EW functional perspective.
- Recommend any EW task organization adjustments.
- Identify key EW decision points.
- Provide EW data for synchronization matrix.
- Recommend EW priority intelligence requirements.
- Identify EW supporting tasks to any branches and sequels.
- Identify potential EW high-value targets.
- Assess EW risks created by telegraphing intentions, allowing time for enemy to mitigate effects, unintended effects of electronic attack, and the impact of asset or capability shortfalls.

4-23. By the conclusion of COA analysis (war-gaming), the EW officer and EW working group members generate or gather the following products and information:

- The EW data for the synchronization matrix.
- The EW portion of the branches and sequels.
- A list of high-value targets related to EW.
- A list of commander's critical information requirements related to EW.
- The risk assessment for EW operations in support of each COA.
- An update to the EW running estimate.

Table 4-1. Sample input to synchronization matrix

TIME/EVENT		H - 8	H - hour	H + 8
Enemy Actions		Enemy monitors movements	Defends from Security Zone	Commits reserve
Decision Points		Launch deep attack		
MANEUVER	1st Brigade	Move on route Paula	Cross line of departure	Seize objective Nick
MANEUVER	2d Brigade	Move on route Mike	Cross line of departure	Seize objective Dave
MANEUVER	3rd Brigade	Move on route Sean		Forward passage of lines with 1st Brigade
MANEUVER	Aviation Brigade	Deep attack on objective Rose at H - 1		
MANEUVER	Division Cavalry		Screen northern flank	
FIRES	Fires Brigade	Preparation fires initiated at H - 5		
FIRES	Air Defense	Weapons hold	Weapons tight	Weapons tight
FIRES	C2W - EA - CNA - Physical Attack - CNE - ES	- Locate enemy ISR on maneuver routes - Deny and disrupt enemy ISR of maneuver routes at H - .5 to H - hour - Disrupt and destroy known enemy C2 nodes and IADS	- Activate CREW systems - Jamming (to disrupt/deny enemy C2 nodes) - Electronic deception - Provide indications and warnings to maneuver brigades	Disrupt and destroy enemy C2 system
C2			Tactical CP with lead brigade	

C2	command and control	EA	electronic attack
C2W	command and control warfare	ES	electronic warfare support
CNA	computer network attack	EW	electronic warfare
CNE	computer network exploitation	H-hour	specific time an operation or exercise begins
CP	command post	IADS	integrated air defense system
CREW	counter radio-controlled improvised explosive device electronic warfare	ISR	intelligence, surveillance, and reconnaissance

Note: This is not complete. Its intent is to show how EW can be integrated into a synchronization matrix.

COURSE OF ACTION COMPARISON

4-24. COA comparison starts with all staff members analyzing and evaluating the advantages and disadvantages of each COA from their perspectives. Staff members present their findings for the others' consideration. Using the evaluation criteria developed during COA analysis, the staff outlines each COA, highlighting its advantages and disadvantages. Comparing the strengths and weaknesses of the COAs identifies their advantages and disadvantages with respect to each other. (See FM 5-0 for further discussion of COA comparison).

4-25. During COA comparison, the EW officer and EW working group members compare COAs based on the EW-related advantages and disadvantages (see center of figure 4-4). Typically, planners use a matrix to assist in the COA comparisons. The EW officer may develop an EW functional matrix to compare the COAs or to use the decision matrix developed by the staff. Regardless of the matrix used, the evaluation criteria developed prior to war-gaming are used to compare the COAs. Normally, the chief of staff or executive officer weights each criterion used for the evaluation based on its relative importance and the commander's guidance. (See FM 5-0 for more information on COA comparison and a sample decision matrix.)

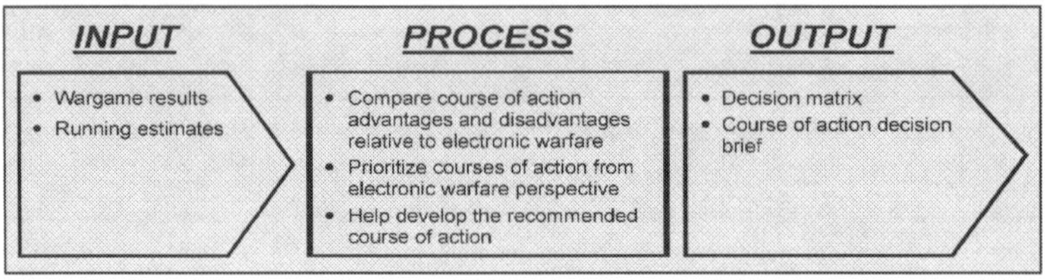

Figure 4-4. Course of action comparison

4-26. By the conclusion of COA comparison, the EW officer and EW working group members generate or gather the following products and information:

- A list of the pros and cons for each COA relative to EW.
- A prioritized list of the COAs from an EW perspective.
- An update to the EW running estimate if required.

COURSE OF ACTION APPROVAL

4-27. The COA approval process has three components. First, the staff recommends a COA, usually in a decision briefing. Second, the commander decides which COA to approve. Lastly, the commander issues the final planning guidance.

4-28. During COA approval, the EW officer supports the development of the COA decision briefing and the development of the warning order as required. If possible, the EW officer attends the COA decision briefing to receive the commander's final planning guidance. If unable to attend the briefing, the EW officer receives the final planning guidance from the G-3 or S-3. The final planning guidance is critical in that it normally provides—

- Refined commander's intent.
- New commander's critical information requirements to support the execution of the chosen COA.
- Risk acceptance.
- Guidance on priorities for the elements of combat power, orders preparation, rehearsal, and preparation.

4-29. After the COA decision has been made, the EW officer and EW working group members generate or gather the following products and information:

- An updated command and control nodal analysis of the enemy relevant to the selected COA.
- Required requests for information to refine the enemy command and control nodal architecture.
- Latest electronic order of battle tailored to the selected COA.
- Any new direction provided in the refined commander's intent.
- A list of any new commander's critical information requirements that can be used in support of EW operations.
- The warning order to assist developing EW operations required to support the operation order or plan.
- Refined input to the initial intelligence, surveillance, and reconnaissance (ISR) plan, including—
  - Any additional specific EW information requirements.
  - Updated potential collection assets for the unit's ISR plan.

ORDERS PRODUCTION

4-30. Orders production consists of the staff preparing the operation order or plan by converting the selected COA into a clear, concise concept of operations. The staff also provides supporting information that enables subordinates to execute and implement risk controls. They do this by coordinating and integrating risk controls into the appropriate paragraphs and graphics of the order.

4-31. During orders production, the EW officer provides the EW operations input for several sections of the operation order or plan. See appendix B for the primary areas for EW operations input within an Army order or plan. The primary areas for EW input in a joint order, if required, also are shown in appendix B. (See CJCSM 3122.03C for the Joint Operation Planning and Execution System format).

DECISIONMAKING IN A TIME-CONSTRAINED ENVIRONMENT

4-32. In a time-constrained environment, the staff might not be able to conduct a detailed MDMP. The staff may choose to abbreviate the process as described in FM 5-0. The abbreviated process uses all seven steps of the MDMP in a shortened and less detailed manner.

4-33. The EW officer and core members of the EW working group meet as a regular part of the unit battle rhythm. However, the EW officer calls unscheduled meetings if situations arise that require time-sensitive planning. Regardless of how much they abbreviate the planning process, the EW officer and supporting members of the EW working group always—

- Update the EW running estimate in terms of assets and capabilities available.
- Update essential EW tasks with the requirements of the commander's intent.
- Coordinate support requests and intelligence requirements with appropriate staff elements and outside agencies.
- Provide EW input to fragmentary orders through the G-3 or S-3 as necessary to drive timely and effective EW operations.
- Deconflict planned EW actions with other uses of the spectrum, such as communications.
- Synchronize electronic attack and EW support actions.
- Synchronize other intelligence collection in support of EW requirements.
- Deconflict EW activities specifically with aviation operations.
- Synchronize EW support to the command and control warfare and information protection information tasks.

THE INTEGRATING PROCESSES AND CONTINUING ACTIVITIES

4-34. Commanders use several integrating processes and continuing activities to synchronize operations throughout the operations process. (See figure 4-5.) The EW officer ensures EW operations are fully synchronized and integrated within these processes and continuing activities. Other staff members supporting the EW working group assist the EW officer. Paragraphs 4-35 through 4-52 outline some key integrating processes and continuing activities. These processes and activities require EW officer involvement throughout the operations process.

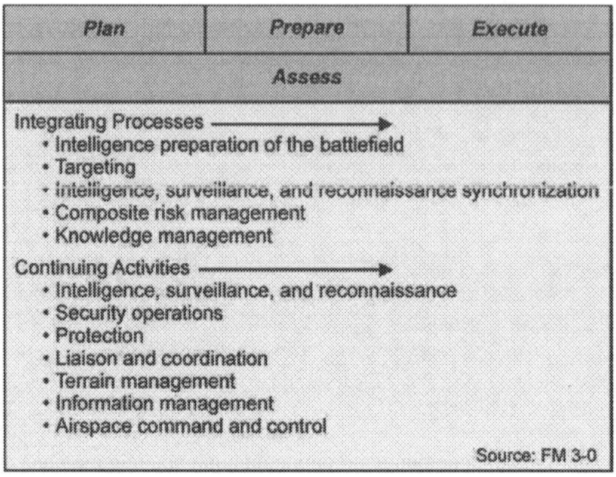

Figure 4-5. Integrating processes and continuing activities

INTELLIGENCE PREPARATION OF THE BATTLEFIELD

4-35. *Intelligence preparation of the battlefield* is the systematic, continuous process of analyzing the threat and environment in a specific geographic area. Intelligence preparation of the battlefield is designed to support the staff estimate and military decisionmaking process. Most intelligence requirements are generated as a result of the intelligence preparation of the battlefield process and its interrelation with the decisionmaking process (FM 34-130). The G-2 or S-2 leads IPB planning with participation by the entire staff. This planning activity is used to define and understand the operational environment and the options it presents to friendly and adversary forces. Only one IPB planning activity exists within each headquarters; all affected staff cells participate. (FM 2-0 provides more information on IPB.) Paragraphs 4-36 through 4-40 discuss how the EW officer and the EW working group support IPB during operations.

4-36. In addition to the input provided to the initial IPB (during step 2 of mission analysis), the EW officer supports IPB throughout the operations process by providing input related to EW operations. (See figure 4-6.) This input includes (but is not limited to) the following EW considerations:

- Evaluating the operational environment from an EW perspective.
- Describing how the effects of the operational environment may impact EW operations.
- Evaluating the threat's capabilities; doctrinal principles; and tactics, techniques, and procedures from an EW perspective.
- Determining threat COAs.

4-37. When evaluating the operational environment from an EW perspective, the EW officer—

- Determines the electromagnetic environment within the defined physical environment:
  - Area of operations.
  - Area of influence.
  - Area of interest.
- Uses electronic databases to identify gaps.
- Identifies adversary fixed EW sites such as EW support and electronic attack sites.
- Identifies airfields and installations that support, operate, or house adversary EW capabilities.
- In coordination with the G-2 or S-2 and G-6 or S-6, helps identify enemy electromagnetic spectrum usage and requirements within the area of operations and area of interest.

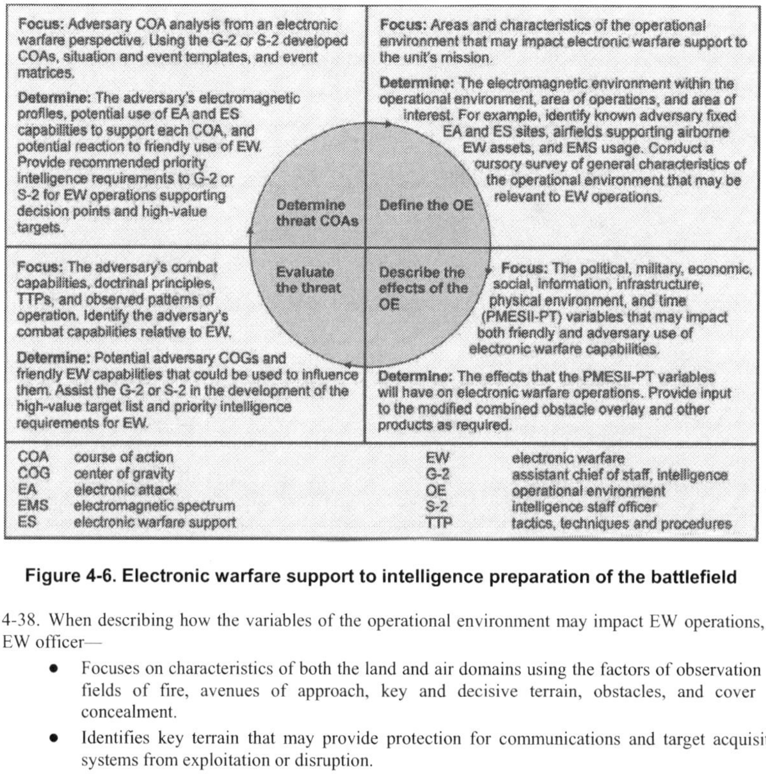

Figure 4-6. Electronic warfare support to intelligence preparation of the battlefield

4-38. When describing how the variables of the operational environment may impact EW operations, the EW officer—

- Focuses on characteristics of both the land and air domains using the factors of observation and fields of fire, avenues of approach, key and decisive terrain, obstacles, and cover and concealment.
- Identifies key terrain that may provide protection for communications and target acquisition systems from exploitation or disruption.
- Identifies how terrain affects line of sight, including effects on both communications and non-communications emitters.
- Evaluates how vegetation affects radio wave absorption and antenna height requirements.
- Locates power lines and their potential to interfere with radio waves.
- Assesses most likely and most dangerous avenues of approach (air, ground) and where EW operations would likely be positioned to support these approaches.

- If operating within urban terrain, considers how the infrastructure—power plants, power grids, structural heights, and communications and media nodes—may restrict or limit EW capabilities.
- Assists the G-2 or S-2 with the development of a modified combined obstacle overlay.
- Determines how weather—visibility, cloud cover, rain, and wind—may affect ground-based and airborne EW operations and capabilities (for example, no-go weather conditions at an airborne EW launch and recovery base).
- Considers all other relevant aspects of the operational environment that affect EW operations, using the operational variables (PMESII-PT—political, military, economic, social, information, infrastructure, physical environment, and time) and mission variables (METT-TC—mission, enemy, terrain and weather, troops and support available, time available, and civil considerations).

4-39. When evaluating enemy capabilities, the EW officer and supporting staff examine doctrinal principles; tactics, techniques and procedures; and observed patterns of operation from an EW perspective. The EW officer—

- Uses the operational variables (PMESII-PT) and mission variables (METT-TC) to help determine the adversary's critical nodes.
- Collects the required data—operational net assessments, electronic order of battle, and electronic databases—to template the command and control critical nodes and the systems required to support and maintain them.
- Assists the G-2 in determining the adversary's EW-related threat characteristics (order of battle) by identifying—
  - Types of communications equipment available.
  - Types of noncommunications emitters.
  - Surveillance and target acquisition assets.
  - Technological sophistication of the threat.
  - Communications network structure.
  - Frequency allocation techniques.
  - Operation schedules.
  - Station identification methods.
  - Measurable characteristics of communications and noncommunications equipment.
  - Command, control, and communications structure of the threat.
  - Tactics from a communication perspective. Examples are how the enemy deploys command, control, and communications assets; whether or not communications systems are remote; and the level of discipline in procedures, communications security, and operations security.
  - Electronic deception capabilities.
  - Reliance on active or passive surveillance systems
  - Electromagnetic profiles of each node.
  - Unique electromagnetic spectrum signatures.
- Assists the G-2 or S-2 in center of gravity analysis. Helps identify the critical system nodes of the center of gravity and determines what aspects of the system should be engaged, exploited, or attacked to modify the system's behavior or to achieve a desired effect.
- Identifies organic and nonorganic EW capabilities available to achieve desired effects on identified high-value targets.
- Submits initial EW-related requests for information that describe the intelligence support required to support EW operations.
- Obtains the high-value target list, threat templates, and initial priority intelligence requirements list to assist in follow-on EW planning.

4-40. When determining adversary COAs, the EW officer—

- Assists the G-2 or S-2 in development of adversary COAs.
- Provides EW input to the situation templates.
- Ensures event templates include EW named areas of interests.
- Assists in providing EW options for target areas of interest.
- Assists in providing EW options to support decision points.
- Provides EW input to the event template and event matrix.

TARGETING

4-41. *Targeting* is the process of selecting and prioritizing targets and matching the appropriate response to them, considering operational requirements and capabilities (JP 3-0). A decide, detect, deliver, and assess methodology is used to direct friendly forces to attack the right target with the right asset at the right time. (See figure 4-7.) Targeting provides an effective method to match the friendly force capabilities against targets. Commander's intent plays a critical role in the targeting process. The targeting working group strives to thoroughly understand the commander's intent to ensure the commander's intended effects on targets are achieved.

4-42. An important part of targeting is identifying potential fratricide situations and performing the coordination measures to manage and control the targeting effort positively. The targeting working group and staff incorporate these measures into the coordinating instructions and appropriate annexes of the operation plans and orders. (FM 6-20-10 has more information on targeting.)

DECIDE		DETECT	
Determine	**Based on**	**Determine**	**Based on**
<u>What (task)</u>: Enemy focused. Determine what EW tasks are essential to the success of the operation (enemy formation or function to influence, and desired targeting effect). <u>Why (purpose)</u>: Friendly focused. Determine the purpose for the use of EA fires (for example, to clear transit routes for maneuvering forces).	• Commanders initial planning guidance • Mission analysis - Specified and implied tasks - Intelligence preparation of the battlefield + target value analysis = enemy courses of action and high-value targets • Commander's intent	<u>Who/Where</u>: Focused on detection. Assets are deployed to detect high-payoff targets. ES collection assets identify and locate targets that can be influenced by EA. Once targets are identified, EA fires can be used to influence the targets based on the identified weaknesses by the target assessment and SIGINT teams.	• COA development • Concept of fires • War-gaming • COA decision • Scheme of fires • High-payoff target list and attack guidance matrix • Reconnaissance and surveillance plan

ASSESS		DELIVER	
Determine	**Based on**	**Determine**	**Based on**
<u>Effect</u>: Identifies whether the intended effect achieved by the EA fires was successful or not.	• Operations plans or orders • EW task execution • Effects of EA fires • BDA • MOE • Target assessment teams • SIGINT team assessment	<u>Who/When</u>: Focused on delivery. Addresses the who and when portion of task (such as the jamming of a designated target and the duration desired).	• COA development • Concept of fires • War-gaming • COA decision • Scheme of fires • High-payoff target list and attack guidance matrix • Reconnaissance and surveillance plan

BDA	battle damage assessment	ES	electronic warfare support	MOE	measures of effectiveness
COA	course of action	EW	electronic warfare	SIGINT	signals intelligence
EA	electronic attack				

Figure 4-7. Electronic warfare in the targeting process

4-43. The EW officer thoroughly integrates electronic attack in the targeting process and integrates electronic attack fires into all appropriate portions of the operation plan, operation order, and other planning products. In support of EW targeting, the EW officer—

- Helps the targeting working group determine electronic attack requirements against specific high-payoff targets and high-value targets.
- Ensures electronic attack can meet the desired effect (in terms of the targeting objective).
- Coordinates with the signals intelligence staff element through the collection manager to satisfy EW support and electronic attack information requirements.
- Prepares the EW tab and the EW portion of the command and control warfare tab to the fires appendix.
- Provides electronic attack mission management through the tactical operations center or joint operations center and the tactical air control party (for airborne electronic attack).
- Provides electronic attack mission management as the jamming control authority for ground or airborne electronic attack when designated.
- Prepares and coordinates the EW annex for operation plans and operation orders.
- Determines and requests theater Army electronic attack support.
- Recommends to the G-3 or S-3 and the fire support coordinator or fire support officer whether to engage a target with electronic attack.
- Expedites electromagnetic interference reports to the targeting working group. (See appendix D for information on electromagnetic interference reporting.)

Decide

4-44. Decide is the first step in the targeting process. This step provides the overall focus for fires, a targeting plan, and some of the priorities for intelligence collection. As part of the staff in the fires cell, the EW officer assists the targeting working group in planning the target priorities for each phase and critical events of the operation. Initially, the targeting working group does not develop electronic attack targets using any special technique or separately from targets for physical destruction. However, as the process continues, these targets are passed through intelligence organizations and further planned using ISR procedures. The planned use of electronic attack is integrated into the standard targeting products (graphic or text-based). Products that involve electronic attack planning may include—

- High-payoff target list.
- Attack guidance matrix.
- Appendix 4 (Electronic Warfare) to Annex P (Information Operations) of the operation order. (At the time this manual was written, this was the current doctrine for operation orders. This appendix will be revised upon publication of the revised FM 5-0.)

Detect

4-45. Based on what the targeting working group identified as high-payoff targets during the decide step, collection assets are then deployed to detect them. The intelligence enterprise pairs assets to targets based on the collection plan and the current threat situation. When conducting electronic attack operations in support of command and control warfare, ISR units perform EW support tasks linked to and working closely with the electronic attack missions. Electronic warfare support units (with support from the target assessment and signals intelligence staff elements) provide the data—location, signal strength, and frequency of the target—to focus electronic attack assets on the intended target. These assets also identify the command and control system vulnerabilities open to attack by electronic attack assets.

Deliver

4-46. Once friendly force capabilities identify, locate, and track the high-payoff targets, the next step in the process is to deliver fires against those targets. Electronic attack assets must satisfy the attack guidance developed during the decide step. Close coordination between those conducting EW support and electronic attack is critical during the engagement. The EW officer facilitates this coordination and ensures electronic attack fires are fully synchronized and deconflicted with other fires. The EW officer remains aware of the potential for unintended effects between adjacent units when conducting electronic attack. The EW officer continually coordinates with adjacent unit EW officers to mitigate and deconflict these effects during cross-boundary operations. Normally, the G-3, S-3, or fire support coordinator provides requirements and guidance for this coordination and synchronization in the attack guidance matrix, intelligence synchronization matrix, spectrum management plan, and the EW input to the operation plan or operation order annexes and appendixes.

Assess

4-47. Once the target as been engaged, the next step is to assess the engagement's effectiveness. This is done through combat assessment, which involves determining the effectiveness of force employment during military operations. It consists of three elements:

- Munitions effects assessment.
- Battle damage assessment.
- Re-attack recommendations.

4-48. The first two elements, munitions effects assessment and battle damage assessment, are used to inform the commander on the effects achieved against targets and target sets. From this information, the G-2 or S-2 continues to analyze the threat's ability to further conduct and sustain combat operations (sometimes articulated in terms of the effects achieved against the threat's centers of gravity). The last element involves the assessment and recommendation whether or not to re-attack the targets.

4-49. The assessment of a jamming mission used against an enemy's command and control system is unlike fires that can be observed visually. The signals intelligence staff element and units executing the electronic attack mission coordinate continuously to assess mission effectiveness. Close coordination between sensor and shooter allows instant feedback on the success or failure of the intended jamming effects. It also can quickly provide the necessary adjustments to produce desired effects.

Intelligence, Surveillance, and Reconnaissance Synchronization

4-50. *Intelligence, surveillance, and reconnaissance synchronization* is the task that accomplishes the following: analyzes information requirements and intelligence gaps; evaluates available assets internal and external to the organization; determines gaps in the use of those assets; recommends intelligence, surveillance, and reconnaissance assets controlled by the organization to collect on the commander's critical information requirements; and submits requests for information for adjacent and higher collection support (FM 3-0). ISR synchronization considers all assets—both internal and external to the organization. It identifies information gaps and the most appropriate assets for collecting information to fill them.

4-51. Planning for ISR operations begins during mission analysis. Although led by the G-3 or S-3, it is supported by the entire staff, subordinate units, and external partners. ISR operations collect, process, store, display, and disseminate information from a multitude of collection sources. The staff thoroughly understands, integrates, and synchronizes the ISR plan across all echelons.

4-52. The EW officer ensures the ISR plan supports the EW-related information requirements determined during the planning process. The EW officer coordinates these requirements with the signals intelligence staff element through the G-2 or S-2.

EMPLOYMENT CONSIDERATIONS

4-53. EW has specific ground-based, airborne, and functional (electronic attack, electronic warfare support, or electronic protection) employment considerations. The EW officer ensures EW-related employment considerations are properly articulated early in the operations process. Each capability employed has certain advantages and disadvantages. The staff plans for all of these before executing EW operations.

Ground-Based Electronic Warfare Considerations

4-54. Ground-based EW capabilities support the commander's scheme of maneuver. Ground-based EW equipment can be employed by a dismounted Soldier or on highly mobile platforms. Due to the short-range nature of tactical signals direction finding, electronic attack assets are normally located in the forward areas of the battlefield, with or near forward units.

4-55. Ground-based EW capabilities have certain advantages. They provide direct support to maneuver units (for example, through counter-radio-controlled improvised-explosive-device EW and communications or sensor jamming). Ground-based EW capabilities support continuous operations and respond quickly to EW requirements of the ground commander. However, to maximize the effectiveness of ground-based EW capabilities, maneuver units must protect EW assets from enemy ground and aviation threats. EW equipment should be as survivable and mobile as the force it supports. Maneuver units must logistically support the EW assets, and supported commanders must clearly identify EW requirements.

4-56. Ground-based EW capabilities have certain limitations. They are vulnerable to enemy attack and can be masked by terrain. They are vulnerable to enemy electromagnetic deceptive measures and electronic protection actions. In addition, they have distance or propagation limitations against enemy electronic systems.

Airborne Electronic Warfare Considerations

4-57. While ground-based and airborne EW planning and execution are similar, they significantly differ in their EW employment time. Airborne EW operations are conducted at much higher speeds and generally have a shorter duration than ground-based operations. Therefore, the timing of airborne EW support requires detailed planning.

4-58. Airborne EW requires the following:

- A clear understanding of the supported commander's EW objectives.
- Detailed planning and integration.
- Ground support facilities.
- Liaisons between the aircrews of the aircraft providing the EW support and the aircrews or ground forces being supported.
- Protection from enemy aircraft and air defense systems.

4-59. Airborne EW capabilities have certain advantages. They can provide direct support to other tactical aviation missions such as suppression of enemy air defenses, destruction of enemy air defenses, and employment of high-speed antiradiation missiles. They can provide extended range over ground-based assets. Airborne EW capabilities can provide greater mobility and flexibility than ground-based assets. In addition, they can support ground-based units in beyond line-of-sight operations.

4-60. The limitations associated with airborne EW capabilities are time-on-station considerations, vulnerability to enemy electronic protection actions, electromagnetic deception techniques, and limited assets (support from nonorganic EW platforms need to be requested).

Electronic Attack Considerations

4-61. Electronic attack includes both offensive and defensive activities. (Chapter 1 provides a full definition of electronic attack). These activities differ in their purpose. Defensive electronic attack protects friendly personnel and equipment or platforms. Offensive electronic attack denies, disrupts, or destroys enemy capability. In either case, certain considerations are involved in planning for employing electronic attack:

- Friendly communications.
- Intelligence collection.
- Other effects.
- Nonhostile local electromagnetic spectrum use.

- Hostile intelligence collection.
- Persistency of effect.

4-62. The EW officer, the G-2 or S-2, the G-3 or S-3, the G-6 or S-6, the spectrum manager, and the G-7 or S-7 coordinate closely to avoid friendly communications interference that can occur when using EW systems on the battlefield. Coordination ensures that electronic attack systems frequencies are properly deconflicted with friendly communications and intelligence systems or that ground maneuver and friendly information tasks are modified accordingly.

4-63. The number of information systems, EW systems, and sensors operating simultaneously on the battlefield makes deconfliction with communications systems a challenge. The EW officer, the G-2 or S-2, the G-6 or S-6, and the spectrum manager plan and rehearse deconfliction procedures to quickly adjust their use of EW or communications systems.

4-64. Electronic attack operations depend on EW support and signals intelligence to provide targeting information and battle damage assessment. However, EW officers must keep in mind that not all intelligence collection is focused on supporting EW. If not properly coordinated with the G-2 or S-2 staff, electronic attack operations may impact intelligence collection by jamming or inadvertently interfering with a particular frequency being used to collect data on the threat, or by jamming a given enemy frequency or system that deprives friendly forces of that means of collecting data. Either can significantly deter intelligence collection efforts and their ability to answer critical information requirements. Coordination between the EW officer, the fire support coordinator, and the G-2 or S-2 is prevents this interference. In situations where a known conflict between the intelligence collection effort and the use of electronic attack exists, the EW working group brings the problem to the G-3 or S-3 for resolution.

4-65. Other forms of effects rely on electromagnetic spectrum. For example, psychological operations may plan to use a given set of frequencies to broadcast messages, or a military deception plan may include the broadcast of friendly force communications. In both examples, the use of electronic attack could unintentionally interfere or disrupt such broadcasts if not properly coordinated. To ensure electronic attack does not negatively impact planned operations, the EW officer coordinates between fires, network operations, and other functional or integrating cells as required.

4-66. Like any other form of electromagnetic radiation, electronic attack can adversely affect local media and communications systems and infrastructure. EW planners consider unintended consequences of EW operations and deconflict these operations with the various functional or integrating cells. For example, friendly jamming could potentially deny the functioning of essential services such as ambulance or fire fighters to a local population. EW officers routinely synchronize electronic attack with the other functional or integrating cells responsible for the information tasks. In this way, they ensure that electronic attack efforts do not cause fratricide or unacceptable collateral damage to their intended effects.

4-67. The potential for hostile intelligence collection also affects electronic attack. A well-equipped enemy can detect friendly EW capabilities and thus gain intelligence on friendly force intentions. For example, the frequencies Army forces jam could indicate where they believe the enemy's capabilities lie. The EW officer and the G-2 or S-2 develop an understanding of the enemy's collection capability. Along with the red team (if available), they determine what the enemy might gain from friendly force use of electronic attack. (A *red team* is an organizational element comprised of trained and educated members that provide an independent capability to fully explore alternatives in plans and operations in the context of the operational environment and from the perspective of adversaries and others. [JP 2-0])

4-68. The effects of jamming only persist as long as the jammer itself is emitting and is in range to affect the target. Normally this time frame is a matter of seconds or minutes, which makes the timing of such missions critical. This is particularly true when jamming is used in direct support of aviation platforms. For example, in a mission that supports suppression of enemy air defense, the time on target and duration of the jamming must account for the speed of attack of the aviation platform. They must also account for the potential reaction time of enemy air defensive countermeasures. The development of directed-energy weapons may change this dynamic in the future. However, at present (aside from antiradiation missiles), the effects of jamming are less persistent than effects achieved by other means.

Electronic Protection Considerations

4-69. Electronic protection is achieved through physical security, communications security measures, system technical capabilities (such as frequency hopping and shielding of electronics), spectrum management, and emission control procedures. The EW officer and EW working group members must consider the following key functions when planning for electronic protection operations:

- Vulnerability analysis and assessment.
- Monitoring and feedback.
- Electronic protection measures and how they affect friendly capabilities.

Vulnerability Analysis and Assessment

4-70. Vulnerability analysis and assessment forms the basis for formulating electronic protection plans. The Defense Information Systems Agency operates the Vulnerability Analysis and Assessment Program, which specifically focuses on automated information systems and can be very useful in this effort.

Monitoring and Feedback

4-71. The National Security Agency monitors communications security. Their programs focus on telecommunications systems using wire and electronic communications. Their programs can support and remediate the command's communications security procedures when required.

Electronic Protection Measures and Their Effect on Friendly Capabilities

4-72. Electronic protection measures include any measure taken to protect the force from hostile electronic attack actions. However, these measures can also limit friendly capabilities or operations. For example, denying frequency usage to counter-radio-controlled improvised-explosive-device EW systems on a given frequency to preserve it for a critical friendly information system could leave friendly forces vulnerable to certain radio-controlled improvised explosive devices. The EW officer and the G-6 or S-6 carefully consider these second-order effects when advising the G-3 or S-3 regarding electronic protection measures.

Electronic Warfare Support Considerations

4-73. The distinction between whether a given asset is performing a signals intelligence or EW support mission is determined by who tasks and controls the assets, what they are tasked to provide, and the purpose for which they are tasked. Operational commanders task assets to conduct EW support for the purpose of immediate threat recognition, targeting, planning the conduct of future operations, and other tactical actions (such as threat avoidance and homing). The EW officer coordinates with the G-2 or S-2 to ensure all EW support needed for planned EW operations is identified and submitted to the G-3 or S-3 for approval by the commander. This ensures that the required collection assets are properly tasked to provide the EW support. In cases where planned electronic attack actions may conflict with the G-2 or S-2 intelligence collection efforts, the G-3, S-3, or commander decides which has priority. The EW officer and the G-2 or S-2 develop a structured process within each echelon for conducting this intelligence gain-loss calculus during mission rehearsal exercises and predeployment work-ups.

Electronic Warfare Reprogramming Considerations

4-74. Electronic warfare reprogramming refers to modifying friendly EW or target sensing systems in response to validated changes in enemy equipment and tactics or the electromagnetic environment. (See paragraph 1-40 for the complete definition.) Reprogramming EW and target sensing system equipment falls under the responsibility of each Service or organization through its respective EW reprogramming support programs. It includes changes to self-defense systems, offensive weapons systems, and intelligence collection systems. During joint operations, swift identification and reprogramming efforts are critical in a rapidly evolving hostile situation. The key consideration for EW reprogramming is joint coordination. Joint coordination of Service reprogramming efforts ensures reprogramming requirements are identified, processed, and implemented consistently by all friendly forces. During joint operations, EW reprogramming coordination and monitoring is the responsibility of the joint force commander's EW staff. (For more information on EW reprogramming, see FM 3-13.10).

SECTION II — ELECTRONIC WARFARE PREPARATION

4-75. *Preparation* consists of activities performed by units to improve their ability to execute an operation. Preparation includes, but is not limited to, plan refinement; rehearsals; intelligence, surveillance, and reconnaissance; coordination; inspections; and movement (FM 3-0). Preparation creates conditions that improve friendly forces' opportunities for success. It facilitates and sustains transitions, including those to branches and sequels.

4-76. During preparation, the EW officer and members of the EW working group focus their actions on the following activities:

- Revising and refining the EW estimate, EW tasks supporting command and control warfare, and EW support to the overall plan.
- Rehearsing the synchronization of EW support to the plan (including integration into the targeting process, request procedures for joint assets, deconfliction procedures, and asset determination and refinement).
- Synchronizing the collection plan and intelligence synchronization matrix with the attack guidance matrix and EW input to the operation plan or order annexes and appendixes.
- Assessing the planned task organization developed to support EW operations, including liaison officers and organic and nonorganic capabilities required by echelon.
- Coordinating procedures with ISR operational elements (such as signals intelligence staff elements).
- Training the supporting staff members of the EW working group during mission rehearsal exercises.
- Completing precombat checks and inspections of EW assets.
- Completing sustainment preparations for EW assets.
- Coordinate with the G-4 or S-4 to develop EW equipment reporting formats.
- Completing briefbacks by subordinate EW working groups on planned EW operations.
- Refining content and format for the EW officer's portion of the battle update assessment and brief.

SECTION III — ELECTRONIC WARFARE EXECUTION

4-77. *Execution* is putting the plan into action by applying combat power to accomplish the mission and using situational understanding to assess progress and make execution and adjustment decisions (FM 3-0). Commanders focus their subordinates on executing the concept of operations by issuing their intent and mission orders.

4-78. During execution, the EW officer and EW working group members—

- Serve as the EW expert for the commander.
- Maintain the running estimate for EW operations.
- Monitor EW operations and recommend adjustments during execution.
- Recommend adjustments to the commander's critical information requirements based on the situation.
- Recommend adjustments to EW-related control measures and procedures.
- Maintain direct liaison with the fires and network operations cells and the command and control warfare working group (if formed) to ensure integration and deconfliction of EW operations.
- Coordinate and manage EW taskings to subordinate units or assets.
- Coordinate requests for nonorganic EW support.

- Continue to assist the targeting working group in target development and recommend targets for attack by electronic attack assets.
- Receive, process, and coordinate subordinate requests for EW support during operations.
- Receive and process immediate support requests for suppression of enemy air defense or EW from joint or multinational forces; coordinate through fire support officer and fire support coordinator with the battlefield coordination detachment and joint or multinational liaisons for support request.
- Coordinate with airspace control section on all suppression of enemy air defense or EW missions.
- Provide input to the overall assessment regarding effectiveness of electronic attack missions.
- Maintain, update, and distribute the status of EW assets.
- Validate and disseminate cease-jamming requests.
- Coordinate and expedite electromagnetic interference reports with the analysis and control element for targeting and the spectrum manager for potential deconfliction.
- Perform jamming control authority function for ground-based EW within the assigned area of operations (when designated by the jamming control authority).

SECTION IV — ELECTRONIC WARFARE ASSESSMENT

4-79. *Assessment* is the continuous monitoring and evaluation of the current situation, particularly the enemy, and progress of an operation (FM 3-0). Commanders, assisted by their staffs, continuously assess the current situation and progress of the operation and compare it with the concept of operations, mission, and commander's intent. Based on their assessment, commanders direct adjustments, ensuring that the operation remains focused on the mission and commander's intent.

4-80. As depicted in figure 4-5 (page 4-10), assessment occurs throughout every operations process activity and includes three major tasks:

- Continuously assessing the enemy's reactions and vulnerabilities.
- Continuously monitoring the situation and progress of the operation towards the commander's desired end state.
- Evaluating the operation against measures of effectiveness and measures of performance.

4-81. The EW officer and supporting members of the EW working group make assessments throughout the operations process. During planning and preparation activities, assessments of EW are made during the MDMP, IPB, targeting, ISR synchronization, and composite risk management integration.

4-82. The EW officer, in conjunction with the G-5 or S-5, helps develop the measures of performance and measures of effectiveness for evaluating EW operations during execution. A *measure of performance* is a criterion used to assess friendly actions that is tied to measuring task accomplishment (JP 3-0). A *measure of effectiveness* is a criterion used to assess changes in system behavior, capability, or operational environment that is tied to measuring the attainment of an end state, achievement of an objective, or creation of an effect (JP 3-0). In the context of EW, an example of a measure of performance is the percentage of known enemy command and control nodes targeted and attacked by electronic attack means (action) versus the number of enemy command and control nodes that were actually destroyed or rendered inoperable for the desired duration (task accomplishment). Measures of effectiveness are used to determine the degree to which an EW action achieved the desired result. This is normally measured through analysis of data collected by both active and passive means. For example, effectiveness is measured by using radar or visual systems to detect changes in enemy weapons flight and trajectory profiles.

4-83. During execution, the EW officer and members of the EW working group participate in combat assessments within the fires cell to determine the effectiveness of electronic attack employment in support of operations. Combat assessment consists of three elements: munitions effects assessment, battle damage assessment, and reattack recommendations. (Paragraphs 4-47 to 4-49 discuss combat assessment.)

SUMMARY

4-84. The EW officer and staff members supporting the EW working group ensure the successful integration of EW capabilities into operations. The EW officer leads the EW integration effort throughout the operations process. The EW officer must be familiar with and participate in the applicable integrating processes and continuing activities discussed within this chapter.

Chapter 5

Coordination, Deconfliction, and Synchronization

Once the commander approves an operation plan or order and preparations are complete, the electronic warfare officer and supporting staff turn to coordinating, deconflicting, and synchronizing the electronic warfare efforts. They ensure electronic warfare actions are carried out as planned or are modified in response to current operations. This chapter discusses major areas and activities that require continuous coordination, deconfliction, and synchronization by the electronic warfare officer and supporting staff of the electronic warfare working groups.

COORDINATION AND DECONFLICTION

5-1. A certain amount of coordination is part of the planning process. However, once a plan is approved and an operation begins, the electronic warfare (EW) staff effort shifts to the coordination and deconfliction necessary to ensure units carry out EW actions as planned or modify actions to respond to the dynamics of the operation.

5-2. The EW officer and members of the EW working group continuously monitor several key areas. These include EW coordination across organizations (higher, lower, and adjacent units), support request coordination, electromagnetic spectrum management, EW asset management, functional coordination between EW subdivisions, EW reprogramming, and EW deconfliction. Normally, EW personnel on watch in the operations center monitor and coordinate activities of these key areas. They alert the EW officer or other EW support personnel to address the required actions.

Coordination Across Organizations

5-3. At the joint level, the information operations division of the J-3 performs EW coordination. The EW section of the information operations staff engages in all EW functions. This section performs peacetime contingency planning, completes day-to-day planning and monitoring of routine theater EW activities, and crisis action planning for contingencies as part of emergent joint operations. The EW section coordinates closely with other appropriate staff sections and other larger joint planning groups as required. (JP 3-13.1 discusses joint EW coordination.)

5-4. In the early stages of contingencies, the joint force commander's EW staff assesses the staffing requirements for planning and execution. This staff also coordinates EW planning and course of action development with the joint force commander's components. Services begin component EW planning and activate their EW working groups per combatant command or Service guidelines. When the scope of a contingency becomes clearer, the command EW officer may request that the joint force commander establish a joint EW coordination cell. If a joint EW coordination cell is formed, it normally requires additional augmentation from the Service or functional components. Depending on the size of the force, EW personnel from the division, corps, or theater are expected to augment the joint EW coordination cell to form a representative EW planning and execution organization. The senior Army organization's staff EW officer anticipates this requirement and prepares to support the augmentation if requested.

5-5. Coordination occurs through established EW working groups from theater level to battalion level. Within Army organizations, the coordination of EW activities occurs both horizontally and vertically. At every level, the staff EW officer ensures the necessary coordination. Normally, coordination of EW activities between the Army and joint force air component commander flows through the battlefield coordination detachment at the joint air operations center. EW staffs at higher echelons monitor EW-related activities and resolve conflicts when necessary.

5-6. Normally the senior Army headquarters (ARFOR) G-3 or S-3 coordinates with external EW organizations, unless direct liaison is authorized at lower echelons. Other components requesting Army EW support coordinate their support requirements with the EW officer located at the ARFOR headquarters or tactical operations center. Often, a liaison from the requesting organization completes these requests. If other Service or functional components have an immediate need for Army EW support, they send the request to the operational fires directorate or fires cell and the senior headquarters EW working group (sometimes referred to as an EW coordination cell) via the Global Command and Control System or Global Command and Control System-Army. In support of external EW coordination, the staff EW officer within the J-3, G-3, or S-3—

- Provides an assessment of EW capabilities to other component operation centers.
- Coordinates preplanned EW operations with other Service components (within prescribed time lines).
- Updates preplanned EW operations in coordination with other components as required.

SUPPORT REQUEST COORDINATION

5-7. Units requesting electronic attack support forward requests to the appropriate EW working group. (See appendix D for the electronic attack request format.) Each EW working group prioritizes the requests and forwards them to the higher headquarters. The commander who owns the capability when the requested support is needed approves the requests. The technical data required to support the execution of the request is passed through EW channels at the appropriate level of classification.

5-8. Electronic warfare support requests are prioritized and passed from the EW working groups through G-2 or S-2 channels and are approved by the commander who owns the capability. New EW support requests are integrated into the intelligence synchronization process. If they are approved, they appear in the intelligence synchronization plan and the unit intelligence, surveillance, and reconnaissance plan. See FMI 2-01 for details on the intelligence synchronization process. The technical data required to support EW support requests passes via signals intelligence channels within the G-2 or S-2 by classified means.

ELECTROMAGNETIC SPECTRUM MANAGEMENT

5-9. The electromagnetic spectrum is a finite resource. Once apportioned, this resource must be managed efficiently to maximize the limited spectrum allocated to support military operations. Electromagnetic spectrum operations aim to enable electronic systems to perform their functions in the intended environment without causing or experiencing unacceptable interference. Electromagnetic spectrum operations deconflict all military, national, and host-nation systems being used in the area of operations, including electronic protection systems, communications systems, sensors, and weapon systems.

5-10. Spectrum management involves planning, coordinating, and managing use of the electromagnetic spectrum through operational, engineering, and administrative procedures. Primarily, it involves determining what specific activities will occur in each part of the available spectrum. For example, some frequencies are assigned to the counter radio-controlled improvised-explosive-device EW systems operating in the area of operations. These frequencies then are deconflicted with ground tactical communications. The spectrum manager ensures all necessary functions that require use of the electromagnetic spectrum have sufficient allocation of that spectrum to accomplish their purpose. Where a conflict (two or more functions require the same portion of the spectrum) exists, the spectrum manager resolves the conflict through direct coordination. Figure 5-1 shows the basic procedures the spectrum manager follows to deconflict spectrum use.

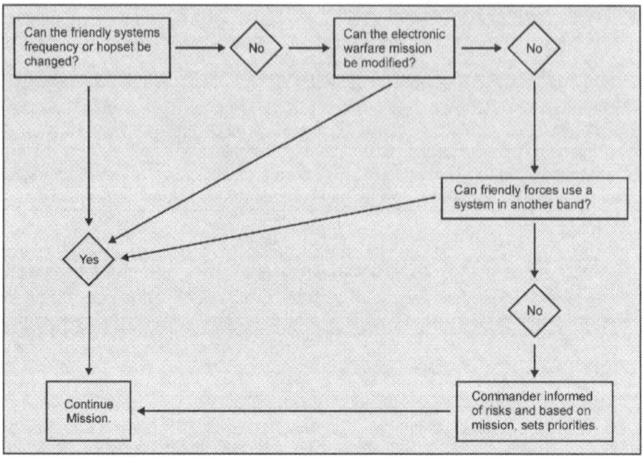

Figure 5-1. Spectrum deconfliction procedures

5-11. The spectrum manager is a member of the G-6 or S-6 section that has staff responsibility for spectrum management in the unit. The spectrum manager is a member of the unit's EW working group. Conflicts regarding spectrum use and allocation that cannot be resolved through direct coordination by the spectrum manager are referred to the G-3 or S-3 for resolution.

Jamming Control Authority

5-12. Depending on the operational situation, an Army headquarters may be designated as the jamming control authority. This authority serves as the senior jamming control authority in the area of operations. It establishes guidance for jamming on behalf of the joint force commander. If designated as the jamming control authority, the senior staff EW officer normally is tasked with the following responsibilities:

- Participating in development of and ensuring compliance with the joint restricted frequency list.
- Validating and approving or denying cease-jamming requests.
- Maintaining situational awareness of all jamming-capable systems in the area of operations.
- Acting as the joint force commander's executive agent for developing EW intelligence gain-or-loss recommendations when electronic attack or electronic warfare support conflicts occur.
- Coordinating jamming requirements with joint force components.
- Investigating unauthorized jamming events and implementing corrective measures.

See JP 3-13.1 for further information on jamming control authority.

Asset Management

5-13. Regardless of echelon, the EW officer monitors and tracks the organization's EW assets and their status. The EW officer makes recommendations to the G-3 or S-3 concerning EW asset allocation and reallocation when required. The EW officer monitors and tracks EW asset status within the EW working group and reports this information to higher echelons via the Army battle command system.

Other Coordinating Actions

5-14. In addition to the functional considerations listed in chapter 4, several coordinating actions must also take place between the EW working groups (at all echelons) and the other planning and execution cells within the headquarters. These actions include—

- Detailed coordination between the EW activities and the intelligence activities supporting an operation.
- Coordination of EW systems reprogramming.
- Coordination with the working groups or cells coordinating the command and control warfare and information protection tasks.

Coordination Between EW Activites and Intelligence Activities

5-15. Most of the intelligence effort, before and during an operation, relies on collection activities targeted against various parts of the electromagnetic spectrum. Electronic warfare support depends on the timely collection, processing, and reporting of intelligence and combat information to alert EW operators and other military activities about intelligence collected in the electromagnetic spectrum. The EW officer and G-2 or S-2 ensure EW collection priorities and EW support collection assets are integrated into a complete intelligence collection plan. This plan ensures that units maximize the use of scarce intelligence and collection assets to support the commander's objectives.

Coordination of EW Systems Reprogramming

5-16. The EW officer and G-2, at division and corps levels, track and coordinate EW systems reprogramming input submitted by lower echelons. This input is then forwarded to the Army Service component command headquarters for submission to the Army Reprogramming Analysis Team. EW officers ensure this input is promptly submitted to ensure urgent reprogramming actions are completed for assigned systems. See FM 3-13.10 for detailed procedures for reprogramming EW and target sensing systems.

Coordination Between EW, Command and Control Warfare, and Information Tasks

5-17. EW working groups coordinate their supporting actions with the elements responsible for the Army information tasks—information engagement, command and control warfare, information protection, operations security, and military deception. Although EW plays a major role in supporting command and control warfare and information protection, it also enhances or provides direct support to other information tasks. For example, enemy radio and television broadcasts can be disrupted or replaced with friendly radio and television messages as part of larger psychological operations in support of information engagement. Electronic deception capabilities can support and enhance an overall military deception operation.

Deconfliction

5-18. Friendly forces depend on electromagnetic energy and the electromagnetic spectrum to sense, process, store, measure, analyze, and communicate information. This dependency creates the potential for significant interference between various friendly systems. Without proper deconfliction, interference could damage friendly capabilities or lead to operational failure. This is especially true with regard to EW systems. EW deconfliction includes—

- Friendly electromagnetic spectrum use for communications and other purposes (such as navigation systems and sensors) with electronic attack activities (such as counter-radio-controlled improvised-explosive-device EW systems).
- Electronic attack activities with electronic warfare support activities (potential electromagnetic interference of collection assets).

- Electronic attack and electronic warfare support activities with information tasks involving electromagnetic emissions (such as counter-radio-controlled improvised-explosive-device EW systems interfering with a psychological-operations radio broadcast).
- Electronic attack activities with host-nation electromagnetic spectrum users (such as commercial broadcasters, emergency first responders, and law enforcement).

5-19. The forum for deconfliction is the unit's EW working group. As such, the specific composition of the working group may expand to include more than the standard staff representation described in chapter 3. Regardless of echelon, to perform its critical deconfliction function, the EW working group retains knowledgeable representation from and ready access to decisionmakers. The EW working group also retains knowledge of and access to higher headquarters assistance and reachback capabilities available (See appendix F for more information).

SYNCHRONIZATION

5-20. EW, particularly in electronic attack, can produce both intended and unintended effects. Therefore, units thoroughly synchronize its use with other forms of fires and with friendly systems operating in the electromagnetic spectrum. Through synchronization, units avoid negative effects such as communications fratricide by jammers. The EW officer ensures all EW activities are integrated into the appropriate sections of plans—fires, information protection, command and control warfare, and military deception plans. This officer also synchronizes EW activities for maximum contribution to the commander's desired effects while preventing EW from inhibiting friendly force capabilities. The primary forum for this synchronization is the unit's EW working group. The EW officer attends the regular targeting meetings in the fires cell and may also participate (perhaps as a standing member) in other functional or integrating cells and working groups. These may include fires, information engagement, network operations, or future operations. The EW officer's participation in these other cells and working groups helps to synchronize EW operations.

SUMMARY

5-21. EW capabilities yield many advantages for the commander. The EW working group's sole purpose is to facilitate the integration, coordination, deconfliction, and synchronization of EW operations to ensure advantages are achieved. This effort requires constant coordination with the unit's other functional cells and working groups. As conflicts are identified during the planning and execution of operations, the EW officer and supporting staff members coordinate solutions to those conflicts within the EW working group.

Chapter 6

Integration with Joint and Multinational Operations

Joint warfare is team warfare. It requires the integrated and synchronized application of all appropriate capabilities. During joint operations, Services work together to accomplish a mission. In multinational operations, forces of two or more nations work together to accomplish a mission. During both joint and multinational operations, forces operate under established organizational frameworks and coordination guidelines. This chapter describes the joint and multinational operational frameworks and guidelines for integrating electronic warfare capabilities.

JOINT ELECTRONIC WARFARE OPERATIONS

6-1. One strength of operating as a joint force is the ability to maximize combat capabilities through unified action. However, the ability to maximize the capabilities of a joint force requires guidelines and an organizational framework that can be used to integrate them effectively. JP 3-13.1 establishes the guidelines and organizational framework for joint electronic warfare (EW) operations.

6-2. Joint task forces are task-organized. Therefore, their composition varies based on the mission. Normally the EW organization within a joint force centers on the—

- Component commands.
- Supporting joint centers.
- Joint force staff.
- Joint force commander's EW staff, joint electronic warfare coordination cell, or information operations (IO) cell.

The supporting centers for EW operations may include the joint operations center, joint intelligence center, Joint Frequency Management Office (JFMO), and joint targeting coordination board.

JOINT FORCE PRINCIPAL STAFF FOR ELECTRONIC WARFARE

6-3. In EW, the principal staff consists of the J-2, J-3, and J-6. The J-2 collects, processes, tailors, and disseminates all-source intelligence for EW. The J-3 has primary staff responsibility for EW activity. This director also plans, coordinates, and integrates joint EW operations with other combat disciplines in the joint task force. Normally, the joint force commander's EW staff or a joint EW coordination cell and an IO cell assist the J-3. The joint force staff network operations director (in the J-6) coordinates electromagnetic spectrum use for information systems with electromagnetic-dependent weapons systems used by the joint force. The IO officer is the principal IO advisor to the J-3. This officer is the lead planner for integrating, coordinating, and executing IO. The command EW officer is the principal EW planner on the J-3 staff. This officer coordinates with the IO cell to integrate EW operations fully with other IO core, supporting, and related capabilities (see JP 3-13.1 for further information)

JOINT FORCE COMMANDER'S ELECTRONIC WARFARE STAFF

6-4. A joint force commander's EW staff supports the joint force commander in planning, coordinating, synchronizing, and integrating joint force EW operations. The joint force commander's EW staff ensures that joint EW capabilities support the joint force commander's objectives. The joint force commander's EW staff is an element within the J-3. It consists of representatives from each component of the joint force.

An EW officer appointed by the J-3 leads this element. The joint force commander's EW staff includes representatives from the J-2 and J-6 to facilitate intelligence support and EW frequency deconfliction.

6-5. On many joint staffs, the intra-staff coordination previously accomplished through a joint force commander's EW staff is now performed by an IO cell or similar organization. An IO cell, if established, coordinates EW activities with other IO activities to maximize effectiveness and prevent mutual interference. If both a joint force commander's EW staff and an IO cell exist, a joint force commander's EW staff representative may be assigned to the IO cell to facilitate coordination. For more information about the organization and procedures of the joint IO cell, see JP 3-13.

Joint Electronic Warfare Coordination Cell

6-6. The decision to form a joint EW coordination cell depends on the anticipated role of EW in an operation. When EW is expected to play a significant role in the joint force commander's mission, a component command's EW coordination organization may be designated as the joint EW coordination cell to handle the EW aspects of the operation. The joint EW coordination cell may be part of the joint force commander's staff, be assigned to the J-3 directorate, or remain within the designated component commander's structure. The joint EW coordination cell plans operational-level EW for the joint force commander. (JP 3-13.1 discusses the joint EW coordination cell in more detail.)

Joint Task Force Component Commands

6-7. Joint task force component commanders exercise operational control of their EW assets. Each component is organized and equipped to perform EW tasks in support of its basic mission and to provide support to the joint force commander's overall objectives. If a component command (Service or functional) is designated to stand up a joint EW coordination cell, it executes the responsibilities and functions outlined in JP 3-13.1.

6-8. A major consideration for standing up a joint EW coordination cell at the component command level is access to a special compartmented information facility to accomplish the cell's required coordination functions. Optimal joint EW coordination cell staffing dictates including special technical operations personnel cleared to coordinate and deconflict special technical operations issues. Special technical operations are associated with the planning and coordination of advanced special programs and the integration of new capabilities into operational units.

6-9. Under current force structure, the special technical operations requirement limits the activation of a joint EW coordination cell to organizations at corps and above levels. Organizations below corps level require significant joint augmentation to meet the special technical operations requirement.

Joint Frequency Management Office

6-10. Joint policy tasks each geographic combatant commander to establish a structure to manage spectrum use and establish procedures that support ongoing operations. This structure must include a JFMO. The JFMO may be assigned from the supported combatant commander's J-6 staff, from a component's staff, or from an external command such as the Joint Spectrum Center. The JFMO coordinates the information systems use of the electromagnetic spectrum, frequency management, and frequency deconfliction. The JFMO develops the frequency management plan and makes recommendations to alleviate mutual interference.

6-11. The G-6 or S-6 coordinates the Army's use of the electromagnetic spectrum, frequency management, and frequency deconfliction with the JFMO through the network operations cell. If established, coordination with the joint spectrum management element is required. (See figure 6-1.)

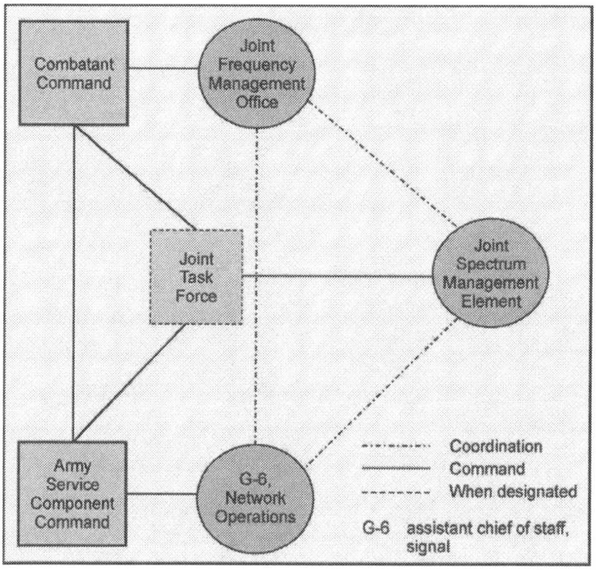

Figure 6-1. Joint frequency management coordination

JOINT INTELLIGENCE CENTER

6-12. The joint intelligence center is the focal point for the intelligence structure supporting the J-2. Directed by the J-2, the joint intelligence center communicates directly with component intelligence agencies and monitors intelligence support to EW operations. This center can adjust intelligence gathering to support EW missions. Within the G-2, EW support requests are coordinated through the requirement cell and then forwarded to the requirements division within the joint intelligence center. (See figure 6-2, page 6-4.)

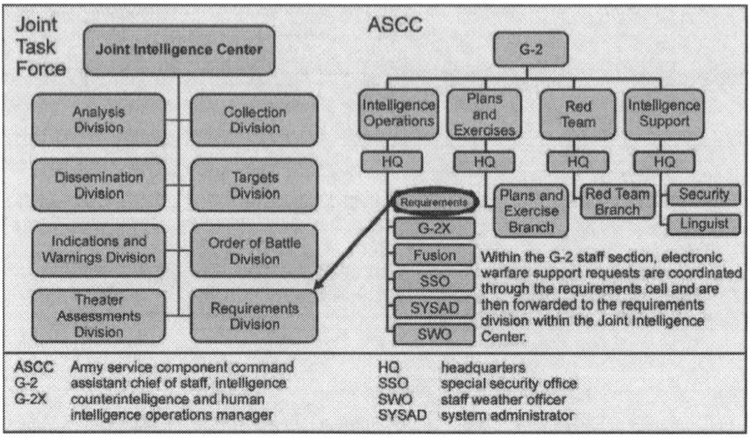

Figure 6-2. Electronic warfare support request coordination

6-13. The composition and focus of each joint intelligence center varies by theater. However, each can perform indications and warning as well as collect, manage, and disseminate current intelligence. Through the joint intelligence center, the ARFOR (Army Service component) headquarters coordinates support from the Air Force, Navy, and Marine Corps and national, interagency, and multinational sources. In addition to its other functions, the joint intelligence center coordinates the acquisition of national intelligence for the joint task force and the combatant command's staff.

Joint Targeting Coordination Board

6-14. The joint targeting coordination board focuses on developing broad targeting priorities and other targeting guidance in accordance with the joint force commander's objectives as they relate operationally. The joint targeting coordination board remains flexible enough to address targeting issues without becoming overly involved in tactical-level decisionmaking. Briefings conducted at the joint targeting coordination board focus on ensuring that intelligence, operations (by all components and applicable staff elements), fires, and maneuver are on track, coordinated, and synchronized. For further information on the joint targeting coordination board, see JP 3-60.

MULTINATIONAL ELECTRONIC WARFARE OPERATIONS

6-15. EW is an integral part of multinational operations (sometimes referred to as combined operations). U.S. planners integrate U.S. and multinational EW capabilities into a single, integrated EW plan. U.S. planners provide multinational forces with information concerning U.S. EW capabilities and provide them EW planning and operational support. However, the planning of multinational force EW is difficult due to security issues, differences in levels of training, language barriers, and terminology and procedural issues. U.S. and North Atlantic Treaty Organization (NATO) EW doctrine provide commonality and a framework for using EW in NATO operations. (See Allied Joint Publication 3.6 for specific information.)

Multinational Force Commander

6-16. The multinational force commander provides guidance for planning and conducting EW operations to the multinational force through the C-3 and the EW coordination cell. The EW coordination cell is located at multinational force headquarters. An IO cell may also be established to coordinate all IO-related activities, including related EW operations.

JOINT OPERATIONS STAFF SECTION

6-17. Within the multinational staff, the joint operations section has primary responsibility for planning and integrating EW activities. A staff EW officer is designated with specific responsibilities. These include integrating multinational augmentees, interpreting or translating EW plans and procedures, coordinating appropriate communications connectivity, and integrating multinational force communications into a joint restricted frequency list.

MULTINATIONAL ELECTRONIC WARFARE COORDINATION CELL

6-18. In multinational operations, the multinational force commander uses an EW coordination cell as the mechanism for coordinating EW resources within the area of operations. This cell is an integral part of the multinational joint force headquarters J-3 staff, at whatever level is appropriate. It provides an effective means of coordinating all EW activities by the multinational force. The multinational force EW coordination cell plans and coordinates all in-theater EW activities in close liaison with the J-2, J-5, and J-6.

ELECTRONIC WARFARE MUTUAL SUPPORT

6-19. Electronic warfare mutual support is the timely exchange of EW information to make the best use of the available resources. It is facilitated by the use of an agreed reference database called the NATO emitter database. Electronic warfare mutual support procedures developed during EW planning include—

- A review of friendly and enemy information data elements that may be exchanged.
- Mechanisms leading to the exchange of data during peace, crisis, and war.
- Development of peacetime exercises to practice the exchange of data.
- Establishment of EW points of contact with adjacent formations and higher and subordinate headquarters for planning purposes, regardless of whether EW resources exist or not.
- Initial acquisition and maintenance of multinational force EW capabilities.
- Exchange of EW liaison teams equipped with appropriate communications.
- Establishment and rehearsal of contingency plans for the exchange of information on friendly and enemy forces.
- Development of communications protocols in accordance with NATO Standardization Agreement (STANAG) 5048.
- Provision of secure, dedicated, and survivable communications.

OTHER CONSIDERATIONS

6-20. EW in multinational operations addresses other considerations. Soldiers must consider—

- Exchange of EW information.
- Exchange of signals intelligence information.
- Exchange of the electronic order of battle.
- Electronic warfare reprogramming.

6-21. Army forces participating in multinational EW operations must exchange EW information with other forces. They must help develop joint information exchange protocols and use those protocols for conducting operations.

6-22. Exchanging signals intelligence information requires care to avoid violating signals intelligence security rules. The policy and relationship between EW and signals intelligence within NATO are set out in NATO Military Committee (MC) 64.

6-23. In peacetime, before forming a multinational force, the exchange of electronic order of battle information is normally achieved under bilateral agreement. During multinational operations, a representative of the joint EW coordination cell, through the theater joint analysis center or the joint intelligence center, ensures the maintenance of an up-to-date electronic order of battle. The inclusion of multinational forces is based on security and information exchange guidelines agreed upon by the participating nations.

6-24. Electronic warfare reprogramming is a national responsibility. However, the joint EW coordination cell remains aware of reprogramming efforts being conducted within the multinational force. FM 3-13.10 guides the Army's reprogramming effort.

SUMMARY

6-25. Every joint or multinational operation is uniquely organized to accomplish the mission. Army EW officers integrate EW forces and capabilities with the organizations and agencies outlined in this chapter. To coordinate Army EW operations with joint and multinational forces, Army EW officers must understand fully the organizational frameworks, policies, and guidelines established for joint and multinational EW operations.

Chapter 7
Electronic Warfare Capabilities

Electronic warfare capabilities consist of high-demand, low-density assets across the Services. Hence, the conduct of electronic warfare operations requires joint interdependence. This complex interdependence extends beyond the traditional Service capabilities. It includes national agencies—such as the Central Intelligence Agency, National Security Agency, and Defense Intelligence Agency—that constantly seek to identify, catalog, and update the electronic order of battle of enemies and adversaries. To support the joint force commander, the subject matter expertise and unique capabilities provided by each Service, agency, and branch or proponent are integrated with all available electronic warfare capabilities.

SERVICE ELECTRONIC WARFARE CAPABILITIES

7-1. Each Service maintains electronic warfare (EW) capabilities to support operational requirements. During operations, the Army is dependent on organic and nonorganic EW capabilities from higher echelons, joint forces, and national agencies. Army EW planners leverage all available EW capabilities to support Army operations. Although not all-inclusive, appendix E provides a listing of current Army, Marine Corps, Navy, and Air Force EW capabilities and references.

EXTERNAL SUPPORT AGENCIES AND ACTIVITIES

7-2. Army EW planners routinely use and receive support from external organizations to assist in planning and integrating EW operations. Support from these organizations may include personnel augmentation, functional area expertise, technical support, and planning support.

BIG CROW PROGRAM OFFICE

7-3. The Big Crow Program Office was established in 1971 to provide testing environments for U.S. military radio frequency sensor, communication, and navigation systems. Today, the Big Crow Program Office provides customers with joint, multifunctional support for testing communications, sensors, information operations, and related weapon systems in support of Department of Defense (DOD), the individual Services, the National Aeronautics and Space Administration, the National Reconnaissance Office, and others. This support includes replicating information operations and EW threat environments as well as providing telemetry recording, technology prototyping, proof-of-concept demonstrations, and information operations and EW training. Big Crow's mission and capabilities now span the electromagnetic spectrum, encompassing EW, telemetry, radar, and electro-optical systems. Mobile and worldwide deployable, the Big Crow Program Office offers a variety of capabilities.

DEFENSE INFORMATION SYSTEMS AGENCY

7-4. The Defense Information Systems Agency is a combat support agency. It plans, develops, fields, operates, and supports command, control, communications, and information systems. These systems serve the President, the Secretary of Defense, the Joint Chiefs of Staff, the combatant commanders, and other DOD components. The Defense Information Systems Agency also operates the Vulnerability Analysis and Assessment Program. This program specifically focuses on automated information systems.

Joint Communications Security Monitor Activity

7-5. The Joint Communications Security Monitor Activity was created in 1993 by a memorandum of agreement between the Services' operations deputies, Directors of the Joint Staff, and the National Security Agency. The Joint Communications Security Monitor Activity monitors (collects, analyzes, and reports) communications security of DOD telecommunications and automated information systems as well as related noncommunications signals. Its purpose is to identify potentially exploitable vulnerabilities and to recommend countermeasures and corrective actions. The Joint Communications Security Monitor Activity supports real world operations, joint exercises, and DOD systems monitoring.

Joint Information Operations Warfare Command

7-6. The Joint Information Operations Warfare Command (JIOWC) was activated in 2006 as a functional component to the United States Strategic Command (USSTRATCOM). JIOWC integrates joint information operations into military plans, exercises, and operations across the spectrum of conflict. It is a valuable resource for commanders during the planning and execution of joint information operations. JIOWC deploys information operations planning teams when the commander of USSTRATCOM approves a request for support. This center delivers tailored, highly skilled support and sophisticated models and simulations to joint commanders and provides information operations expertise in joint exercises and contingency operations.

7-7. JIOWC also fields the Joint Electronic Warfare Center. This center provides specialized expertise in EW. It is an innovation center for existing and emerging EW capabilities and tactics, techniques, and procedures via a network of units, labs, test ranges, and academia. The Joint Electronic Warfare Center also has EW reprogramming oversight responsibilities for the Joint Staff. This oversight includes organizing, managing, and exercising joint aspects of EW reprogramming and facilitating the exchange of joint EW reprogramming data. The actual reprogramming of equipment, however, is a Service responsibility.

Joint Spectrum Center

7-8. The Joint Spectrum Center was activated in 1994 under the direction of the joint staff's J-6. The Joint Spectrum Center assumed all the missions and responsibilities previously performed by the Electromagnetic Compatibility Center plus additional responsibilities. Personnel in the Joint Spectrum Center are experts in spectrum planning, electromagnetic compatibility and vulnerability, electromagnetic environmental effects, information systems, modeling and simulation, operations support, and system acquisition. The Joint Spectrum Center provides complete, spectrum-related services to combatant commanders, Services, and other government agencies. The Joint Spectrum Center deploys teams in support of the combatant commanders and serves as the DOD focal point for supporting spectrum supremacy aspects of information operations. It assists Soldiers in developing and managing the joint restricted frequency list and helps to resolve operational interference and jamming incidents. The Joint Spectrum Center can also provide databases of friendly force command and control systems for use in planning electronic protection. The Joint Spectrum Center is a field office within the Defense Spectrum Organization under the Defense Information Systems Agency.

Joint Warfare Analysis Center

7-9. The Joint Warfare Analysis Center is a Navy-sponsored joint command under the J-3 established in 1994. The Joint Warfare Analysis Center assists the Chairman of the Joint Chiefs of Staff and combatant commanders in preparing and analyzing joint operational plans. It provides analysis of engineering and scientific data and integrates operational analysis with intelligence.

Marine Corps Information Technology and Network Operations Center

7-10. The Marine Corps Information Technology and Network Operations Center is the Marine Corps' enterprise network operations center. The Marine Corps Information Technology and Network Operations Center is the nerve center for the central operational direction and configuration management of the Marine

Corps enterprise network. It is co-located with the Marine Corps forces computer network defense, the component to the joint task force for computer network operations, and the Marine Corps computer incident response team. This relationship provides a strong framework for integrated network management and defense.

National Security Agency

7-11. The National Security Agency/Central Security Service is America's cryptologic organization. This organization protects U.S. government information systems and produces foreign signals intelligence information. Executive Order 12333, 4 December 1981, describes the responsibility of the National Security Agency/Central Security Service in more detail. The resources of National Security Agency/Central Security Service are organized for two national missions:

- The Information Assurance Mission provides the solutions, products, and services, and conducts defensive information operations, to achieve information assurance for information infrastructures critical to U.S. national security interests.
- The Signals Intelligence Mission allows for an effective, unified organization and control of all the foreign signals collection and processing activities of the United States. The National Security Agency is authorized to produce signals intelligence in accordance with objectives, requirements, and priorities established by the Director of National Intelligence in consultation with the President's Foreign Intelligence Advisory Board.

7-12. The Director, National Security Agency is the principal signals intelligence and information security advisor to the Secretary of Defense, Director of National Intelligence, and the Chairman of the Joint Chiefs of Staff. The Director, National Security Agency provides signals intelligence support to combatant commanders and others in accordance with their expressed formal requirements.

SUMMARY

7-13. This chapter and appendix E provide a sampling of available joint and Service EW capabilities, activities, and agencies that support ground force commanders in full spectrum operations. To leverage these capabilities for EW support, Army EW officers acquire a working knowledge of the capabilities available and the procedures for requesting support. Additionally, appendix F provides information on available EW related tools and other resources.

Appendix A
The Electromagnetic Environment

Electromagnetic energy is both a natural and manmade occurrence. This energy, in the form of electromagnetic radiation, consists of oscillating electric and magnetic fields and is propagated at the speed of light. Electromagnetic radiation is measured by the frequency of its wave pattern's repetition within a set unit of time. The standard term for the measurement of electromagnetic radiation is the hertz (Hz), the number of repetitions (cycles) per second. The electromagnetic spectrum refers to the range of frequencies of electromagnetic radiation.

OVERVIEW OF THE ELECTROMAGNETIC ENVIRONMENT

A-1. The electromagnetic environment is the resulting product of the power and time distribution, in various frequency ranges, of radiated or conducted electromagnetic emission levels. Within their intended operational environment, a military force, system, or platform may encounter these emissions while performing tasks during operations. The electromagnetic environment is the sum of—

- Electromagnetic interference.
- Electromagnetic pulse.
- Hazards of electromagnetic radiation to personnel, ordnance, and volatile materials.
- Natural phenomena effects of lightning and precipitation static. (*Precipitation static* is charged precipitation particles that strike antennas and gradually charge the antenna, which ultimately discharges across the insulator, causing a burst of static [JP 3-13.1]).

THE ELECTROMAGNETIC SPECTRUM

A-2. The *electromagnetic spectrum* is the range of frequencies of electromagnetic radiation from zero to infinity. It is divided into 26 alphabetically designated bands (JP 1-02). The spectrum is a continuum of all electromagnetic waves arranged according to frequency and wavelength. The electromagnetic spectrum extends from below the frequencies used for modern radio (at the long-wavelength end) through gamma radiation (at the short-wavelength end). It covers wavelengths from thousands of kilometers to a fraction of the size of an atom. Figure A-1 shows the spectrum regions and wavelength segments associated with the electromagnetic spectrum.

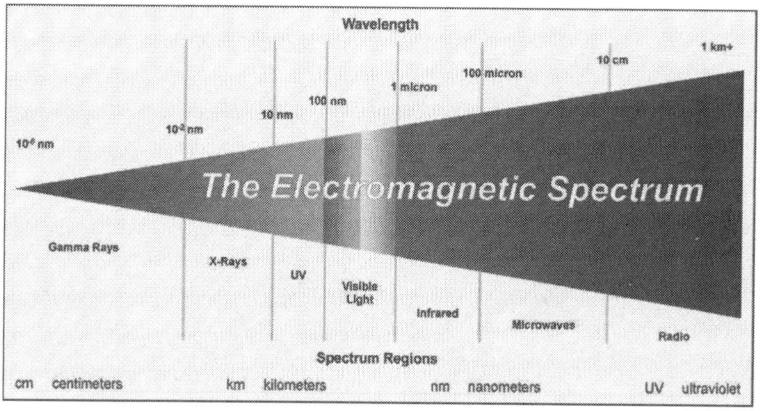

Figure A-1. The electromagnetic spectrum

A-3. Included within the radio and microwave regions of the electromagnetic spectrum are the radio frequency and radar bands. These bands are routinely referred to by their band designators. For example, high frequency radios are HF radios and K-band radars are radars that operate between 18 and 27 gigahertz. Civilian agencies and military forces throughout the world use several different designator systems, which can result in confusion. Table A-1 shows the radio frequency band designators and their associated frequency ranges. It also shows radar band designators, associated frequency ranges, and typical usage. These are standard designations used by the United States.

Table A-1. Radio and radar designators and frequency bands

Radio Frequency Band Designator	Radio Frequency Range	Radar Band Designator*	Frequency Range	Typical Usage
ULF	lower than 3 Hz	VHF	50-330 MHZ	Very long-range surveillance
ELF	3 Hz - 3 kHz	UHF	300-1,000 MHz	Very long-range surveillance
VLF	3 - 30 kHz	L	1-2 Ghz	Long-range surveillance, enroute traffic control
LF	30 - 300 kHz	S	2-4 Ghz	Moderate-range surveillance, terminal traffic control, long-range weather
MF	300 kHz - 3 MHz	C	4-8 Ghz	Long-range tracking, airborne weather
HF	3 - 30 MHz	X	8-12 Ghz	Short-range tracking, missile guidance, mapping, marine radar, airborne intercept
VHF	30 - 300 MHz	K_u	12-18 Ghz	High resolution mapping, satellite altimetry
UHF	300 MHZ - 3 GHz	K	18-27 Ghz	Little use
SHF	3 - 30 GHz	K_a	27-40 Ghz	Very high resolution mapping, airport surveillance
EHF	30 - 300 GHz			
Sub-millimeter	300 Ghz - 1 THz			

EHF extremely high frequency
ELF extremely low frequency
GHz Gigahertz
HF high frequency
Hz hertz
kHz kilohertz
LF low frequency
MF medium frequency
MHz megahertz
SHF super high frequency
THz terahertz
UHF ultra high frequency
ULF ultra low frequency
VHF very high frequency

* Radar band designators relate back to the early development of radar in World War II when the letter designators were used for purposes of secrecy. After the requirement for secrecy was no longer needed, these letter band designators remained.

MILITARY OPERATIONS AND THE ELECTROMAGNETIC ENVIRONMENT

A-4. The impact of the electromagnetic environment upon the operational capability of military forces, equipment, systems, and platforms is referred to as electromagnetic environmental effects. Electromagnetic environmental effects encompass all electromagnetic disciplines, including—

- Electromagnetic compatibility and electromagnetic interference.
- Electromagnetic vulnerability.
- Electromagnetic pulse.
- Electronic protection.
- Hazards of electromagnetic radiation to personnel, ordnance, and volatile materials (such as fuels).
- Natural phenomena effects of lightning and precipitation static.

A-5. *Electromagnetic vulnerability* consists of the characteristics of a system that cause it to suffer a definite degradation (incapability to perform the designated mission) as a result of having been subjected to a certain level of electromagnetic environmental effects (JP 3-13.1). Electronic warfare support plays a key role in identifying the electromagnetic vulnerability of an adversary's electronic equipment and systems. Friendly forces take advantage of these vulnerabilities through electronic warfare operations.

DIRECTED ENERGY

A-6. Directed energy refers to technologies that produce of a beam of concentrated electromagnetic energy or atomic or subatomic particles (see chapter 1). *Directed-energy warfare* is military action involving the use of directed-energy weapons, devices, and countermeasures to either cause direct damage or destruction of enemy equipment, facilities, and personnel, or to determine, exploit, reduce, or prevent hostile use of the electromagnetic spectrum through damage, destruction, and disruption. It also includes actions taken to protect friendly equipment, facilities, and personnel and retain friendly use of the electromagnetic spectrum (JP 1-02). A *directed-energy weapon* is a system using directed energy primarily as a direct means to damage or destroy enemy equipment, facilities, and personnel (JP 1-02). In addition to destructive effects, directed-energy weapons can also support area denial, crowd control, and obscuration.

A-7. The application of directed energy includes lasers, radio-frequency weapons, and particle-beam weapons. As directed-energy weapons evolve, the tactics, techniques, and procedures for their use also evolve to ensure their safe, effective employment. In electronic warfare, most directed-energy applications fit into the category of electronic attack. However, other applications can be categorized as electronic protection or even electronic warfare support. Examples include the following:

- Applications used for electronic attack, which may include—
  - A laser designed to blind or disrupt optical sensors.
  - A millimeter wave directed-energy weapon used for crowd control.
  - A laser-warning receiver designed to initiate a laser countermeasure to defeat a laser weapon.
  - A millimeter wave obscuration system used to disrupt or defeat a millimeter wave system.
  - A device used to counter radio-controlled improvised explosive devices.
- A laser-warning receiver designed solely to detect and analyze a laser signal is used for electronic warfare support.
- A visor or goggle designed to filter out the harmful wavelength of laser light is used for electronic protection.

A-8. As the use of destructive directed-energy weapons grows, Army forces require the capability to collect information on them. Additionally, Army forces require tactics, techniques, and procedures to mitigate directed-energy weapon effects. Currently, the definitions and terms relating to directed energy are articulated within electronic warfare doctrine. As the technologies related to directed energy expand, joint and Army doctrine may discuss employing directed energy under other doctrinal subjects.

Appendix B

Electronic Warfare Input to Operation Plans and Orders

This appendix discusses electronic warfare input to Army and joint plans and orders.

ARMY PLANS AND ORDERS

B-1. This paragraph lists the electronic warfare (EW) information required for Army operation plans and orders. (See figure B-1 on page B-2 for the EW appendix format.) This discussion is based on current doctrine from FM 5-0. When it is republished, FM 5-0 will state where to place EW-related information in the revised plans and orders format. In addition to the appendix 4 (Electronic Warfare) to Annex P (Information Operations), the following components of operation plans and orders may require EW input:

- **Base order or plan**:
  - Sub-subparagraph (2) (Fires) to subparagraph a (Concept of Operations) to paragraph 3 (Execution).
  - Sub-subparagraph (7) (Information Operations) to subparagraph a (Concept of Operations) to paragraph 3 (Execution).
- **Annex D (Fire Support)**:
  - Sub-subparagraph (4) (Electronic Warfare) to subparagraph b (Air Support) to paragraph 3 (Execution)
  - Appendix 1 (Air Support).
- **Annex L (Intelligence, Surveillance, and Reconnaissance)**:
  - Sub-subparagraph (2) (Fires) to subparagraph a (Concept of Operations) to paragraph 3 (Execution).
  - Sub-subparagraph (7) (Information Operations) to subparagraph a (Concept of Operations) to paragraph 3 (Execution).
- **Annex N (Space)**: Sub-subparagraph (10) (Electronic Warfare) to subparagraph b (Space Activities) to paragraph 3 (Execution).
- **Annex P (Information Operations)**:
  - Sub-sub-subparagraph (d) (Electronic Warfare) to sub-subparagraph (8) to subparagraph a (Concept of Support) to paragraph 3 (Execution).
  - Sub-subparagraph (3) (List of Tasks to Electronic Warfare Units) to subparagraph b (Tasks to Subordinate Units) to paragraph 3 (Execution).

[Classification]

Appendix 4 (Electronic Warfare) to Annex P (Information Operations) to OPORD No____

1. SITUATION.

a. Enemy.
- Identify the vulnerabilities of enemy information systems and electronic warfare systems.
- Identify the enemy capability to interfere with accomplishment of the electronic warfare mission.

b. Friendly.
- Identify friendly electronic warfare assets and resources that affect electronic warfare planning by subordinate commanders.
- Identify friendly foreign forces with which subordinate commanders may operate.
- Identify potential conflicts within the friendly electromagnetic spectrum, especially if conducting joint or multinational operations. Identify and de-conflict methods and priority of spectrum distribution.

c. Attachments and detachments.
- List the electronic warfare assets that are attached or detached.
- List the electronic warfare resources available from higher headquarters.

2. MISSION. State how electronic warfare will support the commander's objectives.

3. EXECUTION.

a. Scheme of support. State the electronic warfare tasks.

b. Tasks to subordinate units. Identify the electronic warfare tasks for each unit.

c. Coordinating instructions.
- Identify electronic warfare instructions applicable to two or more units.
- Identify the requirements for the coordination of electronic warfare actions between units.
- Identify the emission control guidance.

4. SERVICE SUPPORT. Identify service support for electronic warfare operations.

5. COMMAND AND SIGNAL.

a. Command.

b. Signal. Identify if any, the special or unusual electronic warfare-related communications requirements.

[Classification]

Figure B-1. Appendix 4 (Electronic Warfare) to annex P (Information Operations) instructions

JOINT PLANS AND ORDERS

B-2. If required to provide EW input to portions of a joint order, the primary areas for input are the following:

- Paragraph 3 (Execution) to appendix 3 (Information Operations) to Annex C (Operations).
- Tab B (Electronic Warfare) to appendix 3 (Information Operations) to Annex C (Operations).

B-3. See CJCSM 3122.03C for the Joint Operations Planning and Execution System format.

Appendix C
Electronic Warfare Running Estimate

This appendix discusses the electronic warfare running estimate. A *running estimate* is a staff section's continuous assessment of current and future operations to determine if the current operation is proceeding according to the commander's intent and if future operations are supportable (FM 3-0).

C-1. The electronic warfare (EW) running estimate is used to support the military decisionmaking process during planning and execution. During planning, the EW running estimate provides an assessment of the supportability of each proposed course of action from an EW perspective. The format of the EW running estimate closely parallels the steps of the military decisionmaking process. It serves as the primary tool for recording the EW officer's assessments, analyses, and recommendations for EW operations. The EW officer and staff in the EW working group are responsible for conducting the analysis and providing recommendations based on the EW running estimate.

C-2. A complete EW running estimate should contain the information necessary to answer any question the commander may pose. If there are gaps in the EW running estimate, the staff identifies the gaps as information requirements and submits them to the intelligence cell. The EW running estimate can form the basis for EW input required in other applicable appendixes and annexes within operation plans and orders. Figure C-1 on page C-2 provides a sample EW running estimate for use during planning.

1. **MISSION.** Show the restated mission resulting from mission analysis.

2. **SITUATION AND CONSIDERATIONS.**

   a. Characteristics of the area of operations.
      - Weather. State how the weather may impact EW operations.
      - Terrain. State how aspects of the terrain may impact EW operations.
      - Civil Considerations. State how rules of engagement and civil emergency responder frequency restrictions may impact EW operations.

   b. Enemy forces. Discuss enemy dispositions, composition, strength, capabilities, and courses of action (COAs) as they affect EW operations. Identify enemy EW vulnerabilities.

   c. Friendly forces.
      - List the current status of the forces EW resources.
      - List the current status of additional EW support resources.
      - Provide a comparison of EW support requirements with available capabilities and recommend solutions for any discrepancies.
      - Identify friendly forces EW vulnerability and recommend solutions.

   d. Assumptions. List any assumptions used that may affect the employment of EW capabilities.

3. **COURSES OF ACTION.**

   a. List the friendly COAs that were wargamed.
   b. List the evaluation criteria identified during the COA analysis.

4. **ANALYSIS.** Analyze each COA using the evaluation criteria identified during COA analysis.

5. **COMPARISION.** Compare each COA. Rank order the COAs for each EW key consideration identified.

6. **RECOMMENDATION AND CONCLUSIONS.** This paragraph translates the "best" course of action (as determined in paragraph 5) into a complete recommendation. It should outline who, what, where, when, how, and why from the EW point of view. It states which course of action can best be supported by friendly EW, and is less vulnerable to enemy EW force capabilities.

   a. Recommend the most supportable COA from an EW perspective.
   b. List any EW related issues, deficiencies and risks and provide recommendations to reduce their impact.

ANNEXES: Include annexes as required. Annexes with pertinent details should be used to the extent practical to support the contents of the estimate. These annexes may be in considerable detail with only the high points included in the body of the estimate. Annexes should add depth to the contents of the estimate, but should not be used as a substitute for key points that should be included in the bodv of the estimate.

Figure C-1. Example of an electronic warfare running estimate

C-3. Once the commander approves the order, the EW running estimate is used to inform current and future operations. During execution the EW running estimate is used to help determine if current EW operations are proceeding according to plan and if future EW operations are supportable. Figure C-2, page C-3, shows a sample of the information that might be used to update the EW running estimate during execution. The EW officer and supporting staff members within the EW working group produce and update the running estimate.

Current operation order and fragmentary orders

- Define the battlefield environment. Focus on the aspects of the terrain and weather that could assist or enable electronic warfare operations from both a friendly and threat viewpoint.
  - Maintain updated weather and terrain data.
  - Locate terrain for communications and non-communications sites, line of sight.
  - Identify aspects of terrain and weather that may have an impact on the electromagnetic spectrum.
- Define the threat.
  - Communications systems, including threat radio nets and network nodes.
  - Noncommunications emitters.
  - Electronic support systems.
  - Electronic attack systems.
- Identify host-nation use of the electromagnetic spectrum (restricted frequencies such as government, industry, and emergency responders).
- Identify friendly capabilities, shortfalls and readiness.
  - Electronic attack capabilities and status (joint and Army).
  - Location and availability of organic friendly electronic warfare capabilities (such as Prophets and counter-radio-controlled IED EW systems.
  - Electronic warfare vulnerabilities.
  - Equipment updates, both hardware and software.
- Identify enemy capabilities, shortfalls and readiness.
  - Electronic warfare capabilities and status (if known).
  - Electronic warfare vulnerabilities.
  - Electronic order of battle.

Electronic Warfare Target Folder

- Describe the targeted capability, its associated vulnerabilities, and the friendly capabilities used to engage them.
- Maintain updated high-value target and high-payoff target lists.
- Develop a prioritized target list based on high-value targets and high-payoff targets.
- EW target folders are split between traditional and asymmetric targets.
  - Traditional targets might include integrated air defense systems, communications nodes, and radar facilities. Traditional targets are normally fixed or less mobile than asymmetric targets and are easier to develop.
  - Asymmetric targets might include individual cell phones, radio-controlled improvised explosive devices, global positioning systems and wireless networks. Asymmetric targets can be either stationary or mobile and are typically harder to develop than traditional targets during the targeting phase.

Figure C-2. Sample update information to the electronic warfare running estimate

Appendix D
Electronic Warfare-Related Reports and Messages

This appendix provides information and references for electronic warfare and electronic warfare-related reports and message formats.

MESSAGES AND SUMMARIES

D-1. The following messages and summaries are associated with the planning, synchronization, deconfliction, and assessment of EW operations.

ELECTRONIC ATTACK DATA MESSAGE

D-2. An electronic attack data message reports an electronic attack strobe from an affected or detecting unit's position to an aircraft emitting an electronic attack. It is used to determine the location of a hostile or unknown aircraft emitting an electronic attack. The detecting unit reports its detection to all units using a given network when the data link is degraded or not operational.

D-3. Upon receipt of several messages, the source of enemy electronic attack can be determined by comparing lines of bearing from the different origins (triangulation).

D-4. See FM 6-99.2, page 83, for the format.

ELECTRONIC ATTACK REQUEST FORMAT

D-5. Electronic fires fall within three categories: preplanned, preplanned on-call, and immediate. Requesting airborne electronic attack support for ground operations is similar to requesting close air support. Requests for an electronic attack are sent via the normal joint air request process. Requesters use either a joint tactical air strike request or joint tactical air support request. (See FM 3-09.32 for a sample.) A theater-specific electronic attack request format may complement a joint tactical air strike request.

D-6. When submitting the request, the following information must be provided in the remarks section (section 8):

- Target location.
- Prioritized target description and jam frequencies.
- Time on target (window).
- Joint terminal attack controller.
- Jamming control authority call sign and frequency.
- Friendly force disposition (for example, troop movement route).
- Friendly frequency restrictions.
- Remarks.

ELECTRONIC WARFARE FREQUENCY DECONFLICTION MESSAGE

D-7. An EW frequency deconfliction message promulgates a list of protected, guarded, and taboo frequencies. This list allows friendly forces to use the frequency spectrum without adverse impact from friendly electronic attack. (See FM 6-99.2, page 86, for the format.)

ELECTRONIC WARFARE MISSION SUMMARY

D-8. The EW mission summary summarizes significant EW missions and reports the status of offensive EW assets. EW and electronic-attack-capable surface and air units use it to provide information on EW operations. Service components use it to report significant events for subsequent analysis. (See FM 6-99.2, page 87, for the format.)

ELECTRONIC WARFARE REQUESTING TASKING MESSAGE

D-9. Joint task force commanders use the electronic warfare requesting tasking message to task component commanders to perform EW operations in support of the joint EW plan and to support component EW operations. Component commanders use this message to request EW support from sources outside their command.

JOINT TACTICAL AIR STRIKE REQUEST OR JOINT TACTICAL AIR SUPPORT REQUEST

D-10. Use a joint tactical air strike request or joint tactical air support request to request electronic attack. These requests require the information listed in paragraph D-6. Organizations without an automated capability submit these requests using DD Form 1972 (Joint Tactical Air Strike Request). See JP 3-09.3 and FM 3-09.32 for more information.

JOINT SPECTRUM INTERFERENCE RESOLUTION

D-11. The joint spectrum interference resolution program replaced the DOD meaconing, intrusion, jamming, and interference program in June, 1992. Follow guidance in CJCSI 3320.02C to report incidents of spectrum interference.

JOINT RESTRICTED FREQUENCY LIST

D-12. Operational, intelligence, and support elements use the joint restricted frequency list to identify the level of protection desired for various networks and frequencies. The list should be limited to the minimum number of frequencies necessary for friendly forces to accomplish objectives.

D-13. See Annex A to appendix B to JP 3-13.1 for the joint restricted frequency list format. The format is used by the joint automated communications-electronics operations instruction system. The format is unclassified but should show the proper classification of each paragraph when filled in. (See CJCSI 3320.01B and JP 3-13.1 for additional information.)

COUNTER-IMPROVISED-EXPLOSIVE-DEVICE ACTIVITIES

D-14. Certain reports and references are associated with counter-improvised-explosive-device activities. Most of these reports include information pertinent to counter-radio-controlled improvised-explosive-device EW activities. EW working groups have the responsibility to monitor these reports to assess planned counter-radio-controlled improvised-explosive-device EW operations and to support future operations. These reports typically use formats established in FM 6-99.2 modified to include improvised explosive device considerations and current operations. See GTA 90-10-046 for examples of reports and references applicable to counter-radio-controlled improvised-explosive-device EW operations.

Appendix E
Army and Joint Electronic Warfare Capabilities

This appendix provides information on Army and other Service electronic warfare capabilities. It is not an all-inclusive list. Due to the evolving nature of electronic warfare equipment and systems, this information is perishable and should be augmented, updated, and maintained by the unit electronic warfare officer.

ARMY

E-1. The Army is currently expanding its electronic warfare (EW) capability. It maintains several EW systems in its inventory. Currently, all units whose sole purpose is to conduct EW operations are assigned to 1st Information Operations Command. When requested, these capabilities are provided to combatant commands for employment at corps and lower echelons.

COUNTER-RADIO-CONTROLLED IMPROVISED-EXPLOSIVE-DEVICE EW SYSTEMS

E-2. Counter-radio-controlled improvised-explosive-device EW systems form a family of electronic attack systems. Army forces use these systems to prevent improvised explosive device detonation by radio frequency energy. The Army maintains both a mounted and dismounted counter-radio-controlled improvised-explosive-device EW capability to protect personnel and equipment. For a detailed description of these systems, see appendix F.

AIRCRAFT SURVIVABILITY EQUIPMENT

E-3. Aircraft survivability equipment aims to reduce aircraft vulnerability, thus allowing aircrews to accomplish their immediate mission and survive. Army aviation maintains a suite of aircraft survivability equipment that provides protection against electronic attack. This protection can include radio frequency warning and countermeasures systems, a common missile warning system, information requirement countermeasures systems, and laser detection and countermeasure systems. For a detailed description of aircraft survivability equipment EW-related systems, see appendix F.

INTELLIGENCE SYSTEMS

E-4. The intelligence community maintains many systems that provide data for use in EW operations. Signals intelligence systems provide most of this required data. These assets are dual use. Usually the data collected is categorized as signals intelligence. It is maintained within sensitive compartmented information channels and governed by the National Security Agency/Central Security Service. The data sometimes support EW or, more specifically, electronic warfare support. Paragraphs E-5 through E-7 illustrate some intelligence systems that (when tasked) can provide electronic warfare support data to support electronic attack and electronic protection actions. For a detailed description of other intelligence and EW-support-related systems, see appendix F.

Guardrail Common Sensor

E-5. The Guardrail common sensor is a corps-level airborne signals intelligence collection and location system. (See figure E-1.) It provides tactical commanders with near real-time targeting information. Key features include the following: integrated communications intelligence and electronic intelligence reporting, enhanced signal classification and recognition, near real-time direction finding, precision emitter location, and an advanced integrated aircraft cockpit. Preplanned product improvements include frequency extension, computer-assisted online sensor management, upgraded data links, and the capability to exploit a wider range of signals. The Guardrail common sensor shares technology with the ground-based common sensor, airborne reconnaissance-low, and other joint systems.

Figure E-1. Guardrail common sensor

Aerial Common Sensor

E-6. The aerial common sensor is the Army's programmed airborne intelligence, surveillance, and reconnaissance system. (See figure E-2.) It will replace the current RC-7 airborne reconnaissance-low and Guardrail common sensor programs. The aerial common sensor uses the operational and technical legacies of the airborne reconnaissance-low and Guardrail common sensor systems as well as some technological improvements. This sensor will then provide a single, effective, and supportable multiple-intelligence system for the Army. The aerial common sensor will include a full multiple-intelligence capability, including carrying signals intelligence payloads, electro-optic and infrared sensors, radar payloads, and hyperspectral sensors.

Figure E-2. Aerial common sensor (concept)

Prophet

E-7. The Prophet system is the division, brigade combat team, and armored cavalry regiment principal ground tactical signals intelligence and EW system. (See figure E-3.) Prophet systems will also be assigned to the technical collection battalion of battlefield surveillance brigades. Prophet detects, identifies, and locates enemy electronic emitters. It provides enhanced situational awareness and actionable 24-hour information within the unit's area of operations. Prophet consists of a vehicular signals intelligence receiver mounted on a high mobility multipurpose wheeled vehicle, plus a dismounted-Soldier-portable version. The dismounted Soldier portable version is used for airborne insertion or early entry to support rapid reaction contingency and antiterrorist operations. Future Prophet systems are planned to include an electronic attack capability.

Figure E-3. Prophet (vehicle-mounted)

MARINE CORPS

E-8. The Marine Corps has two types of EW units: radio battalions (often called RADBNs), and Marine tactical EW squadrons (referred to as VMAQs). Paragraphs E-9 through E-24 discuss the units' missions, their primary tasks, and capabilities currently being employed. (For further information on the Marine Corps EW units and systems, see MCWP 2-22.)

RADIO BATTALION

E-9. Radio battalions are the Marine Corps' tactical level ground-based EW units. During operations, teams from radio battalions are most often attached to the command element (or senior headquarters) of Marine expeditionary units. Each radio battalion has the following mission, tasks, and equipment.

Mission and Tasks

E-10. The mission of the radio battalion is to provide communications security monitoring, tactical signals intelligence, EW, and special intelligence communication support to the Marine air-ground task force (MAGTF). The radio battalion's tasks include—

- Executing interception; radio direction finding; recording and analysis of communications and noncommunications signals; and signals intelligence processing, analysis, production, and reporting.
- Conducting EW against enemy or adversary communications.
- Helping protect MAGTF communications from enemy exploitation by conducting communications security monitoring, analysis, and reporting on friendly force communications.
- Providing special intelligence communications support and cryptographic guard (personnel and terminal equipment) in support of the MAGTF command element. Normally, the communications unit supporting the MAGTF command element provides communications connectivity for special intelligence communications.
- Providing task-organized detachments to MAGTFs with designated signals intelligence, EW, special intelligence communication, and other required capabilities.
- Exercising technical control and direction over MAGTF signals intelligence and EW operations.
- Providing radio reconnaissance teams with specialized insertion and extraction capabilities (such as combat rubber raiding craft, fast rope, rappel, helocast, and static-line parachute) for specified signals intelligence and limited electronic attack support during advance force, preassault, or deep postassault operations.
- Coordinating technical signals intelligence requirements and exchanging technical information and material with national, combatant command, joint, and other signals intelligence units.
- Providing intermediate, third, and fourth echelon maintenance of the radio battalion's signals intelligence and EW equipment.

Equipment

E-11. The following illustrate EW capabilities a radio battalion uses to accomplish the mission and perform the tasks in support of the MAGTF:

AN/ULQ-19(V)2 Electronic Attack Set

E-12. The AN/ULQ-19(V)2 electronic attack set allows operators to conduct spot or sweep jamming of single-channel voice or data signals. To provide the required jamming, the system must be employed and operated from a location with an unobstructed signal line of sight to the target enemy's communications transceiver.

AN/MLQ-36 Mobile Electronic Warfare Support System

E-13. The AN/MLQ-36 mobile electronic warfare support system provides a multifunctional capability that gives signals intelligence and EW operators limited armor protection. This equipment can provide signals intelligence and EW support to highly mobile mechanized and military operations in urban terrain where maneuver or armor protection is critical. This system is installed in a logistic variant of the Marine Corps's light armored vehicle. It consists of the following:

- Signals intercept system.
- Radio direction finding system.
- Electronic attack system.
- Secure communication system.
- Intercom system.

AN/MLQ-36A Mobile Electronic Warfare Support System (Product Improved)

E-14. The product-improved AN/MLQ-36A mobile electronic warfare support system (sometimes called the AN/MLQ-36A MEWSS PIP) is an advanced signals intelligence and EW system integrated into the Marine Corps's light armored vehicle. (See figure E-4.) This system replaces the equipment in the AN/MLQ-36.

E-15. The AN/AMLQ-36A has the following capabilities:

- Detect and evaluate enemy communications emissions.
- Detect and categorize enemy noncommunications emissions (such as battlefield radars).
- Determine lines of bearing.
- Degrade enemy tactical radio communications.

When mission-configured and working cooperatively with other AN/MLQ-36As, the system can provide precision location of battlefield emitters.

E-16. This system and its future enhancements will provide the capability to exploit new and sophisticated enemy electronic emissions and conduct electronic attack in support of existing and planned national, combatant command, fleet, and MAGTF signals intelligence and EW operations.

Figure E-4. AN/MLQ-36A mobile electronic warfare support system

Marine Tactical Electronic Warfare Squadron

E-17. Marine tactical electronic warfare squadrons are the Marine Corps's airborne tactical EW units. Each squadron has the following mission, tasks, and capabilities.

Mission and Tasks

E-18. The mission of the electronic warfare squadron is to provide EW support to the MAGTF and other designated forces. The squadron conducts tactical jamming to prevent, delay, or disrupt the enemy's ability to use the following kinds of radars: early warning, acquisition, fire or missile control, counterfire, and battlefield surveillance. Tactical jamming also denies and degrades enemy communication capabilities. The squadron conducts electronic surveillance operations to maintain electronic orders of battle. These include both selected emitter parameters and nonfriendly emitter locations. The squadron also provides threat warnings for friendly aircraft, ships, and ground units. Squadron tasks include—

- Providing airborne electronic attack and EW support to the aviation combat element and other designated operations by intercepting, recording, and jamming threat communications and noncommunications emitters.
- Processing, analyzing, and producing routine and time-sensitive electronic intelligence reports for updating and maintaining enemy electronic order of battle.

- Providing liaison personnel to higher staffs to assist in squadron employment planning.
- Providing an air EW liaison officer to the MAGTF EW coordination cell.
- Conducting electronic attack operations for electronic protection training of MAGTF units.

E-19. The squadron's EW division supports EA-6B Prowler tactical missions with intelligence, the tactical electronic reconnaissance processing and evaluation system (TERPES), and the joint mission planning system. All systems support premission planning and postmission processing of collected data, and production of pertinent intelligence reports. Working with squadron intelligence, these systems provide required electronic intelligence and electronic order of battle intelligence products to the aviation combat element, MAGTF, and other requesting agencies.

Equipment

E-20. Marine tactical electronic warfare squadrons maintain the following equipment:

- EA-6B Prowler.
- Joint mission planning system.
- Tactical electronic reconnaissance processing and evaluation system.

EA-6B Prowler

E-21. The EA-6B Prowler is a subsonic, all-weather, carrier-capable aircraft. (See figure E-5.) The crew consists of one pilot and three electronic countermeasure officers. The EA-6B has two primary missions. One is collecting and processing designated threat signals of interest for jamming and subsequent processing, analysis, and intelligence reporting. The other is employing the AGM-88 high-speed antiradiation missile against designated targets. The EA-6B's AN/ALQ-99 tactical jamming system incorporates receivers for the reception of emitted signals and external jamming pods for the transmission of energy to jam targeted radars (principally those associated with enemy air defense radars and associated command and control). In addition to the AN/ALQ-99, the EA-6B also employs the USQ-113 communications jammer to collect, record, and disrupt threat communications.

Figure E-5. EA-6B Prowler

Joint Mission Planning System

E-22. The joint mission planning system helps the EA-6B aircrew plan and optimize receivers, jammers, and high-speed antiradiation missiles. This system allows an operator to—

- Maintain area of operations emitter listings.
- Edit emitter parameters.
- Develop mission-specific geographic data and electronic order of battle to—
  - Tailor or create high-speed antiradiation missile direct attack libraries, or manually modify entries or new threat cards.
  - Plan target selection.
- Perform postflight mission analysis to—
  - Identify electronic emitters using various electronic parameter databases and electronic intelligence analytical techniques.
  - Localize emitters by coordinates with a certain circular error of probability for each site.
  - Correlate new information with existing data.
  - Gather postflight high-speed antiradiation missile information. This information includes aircraft launch parameters, predicted seeker footprint, and the onboard system detection of a targeted signal at impact.

AN/TSQ-90 Tactical Electronic Reconnaissance Processing and Evaluation System

E-23. The TERPES (AN/TSQ-90) is an air and land transportable, single-shelter electronic intelligence processing and correlation system. Each of the four Marine tactical electronic warfare squadrons includes a TERPES section.

E-24. A TERPES section consists of Marines, equipment, and software. The section identifies and locates enemy radar emitters from data collected by EA-6B aircraft and those received from other intelligence sources. It processes and disseminates EW data rapidly to MAGTF and other intelligence centers and provides mission planning and briefing support. Section support areas include operational support, intelligence analysis support, data fusion, fusion processing, and intelligence reporting. The section provides the following operational support:

- Translates machine-readable, airborne-collected, digital data into human- and machine-readable reports (such as paper, magnetic tape, secure voice, plots, and overlays).
- Receives and processes EA-6B mission tapes.
- Accepts, correlates, and identifies electronic emitter data from semiautomatic or automatic collection systems using various electronic parameter databases and various analysis techniques.
- Provides tactical jamming analysis.

AIR FORCE

E-25. The Air Force has two primary platforms that provide EW capability: the EC-130H Compass Call and RC-135V/W Rivet Joint. (For further information on Air Force EW equipment, see AFDD 2-5.1.)

EC-130H COMPASS CALL

E-26. The EC-130H Compass Call is an airborne tactical weapon system. (See figure E-6.) Paragraphs E-27 through E-31 discuss the EC-130H missions, primary tasks, and capabilities.

Mission and Tasks

E-27. The EC-130H's mission is to disrupt enemy command and control information systems and limit the coordination essential for force management. The EC-130H's primary task is to employ offensive counterinformation and electronic attack capabilities in support of U.S. and multinational tactical air, surface, and special operations forces.

Figure E-6. EC-130H Compass Call

Capabilities

E-28. The EC-130H is designed to deny, degrade, and disrupt adversary command and control information systems. This includes denial and disruption of enemy surveillance radars; denial and disruption of hostile communications being used in support of enemy ground, air, or maritime operations; and denial and disruption of many modern commercial communication signals that an adversary might employ.

Compass Call During Operation Iraqi Freedom

During Operation Iraqi Freedom, much speculation appeared in the press about why Iraqi forces failed to ignite the oil facilities they had wired for destruction. During the coalition's seizure of Al Faw, Compass Call disrupted the Iraqi regime's control of its troops by jamming its communications. Instead of receiving orders to detonate the oil terminals, Iraqi troops heard only the ratcheting static of Compass Call jamming until coalition ground troops had secured the area. In addition to the conquest of the Al Faw Peninsula, successful military operations supported by Compass Call in Operation Iraqi Freedom included the seizure of four airfields; two successful prisoner of war rescues; and the ground offensive from Basrah to Nasariyah, Najaf, Baghdad, and Tikrit. In all these instances, Compass Call jamming prevented a trained, experienced enemy from coordinating actions against coalition forces.

"EC-130H Compass Call: A textbook example of Joint Force integration at its best", Electronic Warfare Working Group, U.S. House of Representatives, Issue Brief #17, 11 Mar 2004. (Available at http://www.house.gov/pitts/initiatives/ew/Library/Briefs/brief17.htm)

RC-135V/W Rivet Joint

E-29. Paragraphs E-30 through E-31 discuss the missions, primary tasks, and capabilities of the RC-135V platforms.

Mission and Tasks

E-30. The RC-135V/W Rivet Joint is a combatant-command-level surveillance asset that responds to national-level taskings. (See figure E-7.) Its mission is to support national consumers, combatant commanders, and combat forces with direct, near real-time reconnaissance information and electronic warfare support. It collects, analyzes, reports, and exploits information from enemy command and control information systems. During most contingencies, it deploys to the theater of operations with the airborne elements of the theater air control system.

RC-135V/W Rivet Joint, U.S. Air Force Fact Sheet, September, 2007.

Figure E-7. RC-135V/W Rivet Joint

Capabilities

E-31. The RC-135V/W is equipped with an extensive array of sophisticated intelligence gathering equipment that enables monitoring of enemy electronic activity. The aircraft is integrated into the theater air control system via data links and voice (as required). Refined intelligence data can be transferred from Rivet Joint to an Airborne Warning and Control System platform through the tactical digital information link. Alternatively, this data can be placed into intelligence channels via satellite and the tactical information broadcast service (a near real-time combatant command information broadcast). The aircraft has secure ultrahigh frequency, very high frequency, and high frequency (commonly known as UHF, VHF, and HF respectively) as well as satellite communications. It can be refueled in the air.

NAVY

E-32. The Navy's primary airborne EW platforms are the EA-6B Prowler and its planned replacement, the E/A-18G Growler. E/A-18G fielding is scheduled to begin in 2009 and is scheduled to replace the Navy's carrierborne EA-6B aircraft. The Navy also maintains both surface and subsurface EW shipboard systems for offensive and defensive missions in support of the fleet. (For further information on Navy missions and equipment, see NWP 3-13.)

EA-6B PROWLER

E-33. Paragraphs E-34 through E-39 discuss the missions, primary tasks, and capabilities of the Navy's EA-6B Prowler platforms. (See figure E-8.)

Figure E-8. Navy EA-6B Prowler

Mission and Tasks

E-34. The mission of the Navy's EA-6B Prowler is to ensure survivability of U.S. and multinational forces through suppression of enemy air defenses (using the radar-jamming AN/ALQ-99 tactical jamming system), lethal suppression (using the AGM-88 high-speed antiradiation missile), and communications jamming (using the USQ-113 radio countermeasures set). Prowlers have supported U.S. and multinational forces operating from various expeditionary sites throughout the world while maintaining full presence on all Navy aircraft carriers.

Capabilities

E-35. The Navy's EA-6B Prowlers are outfitted with either the improved capability II or improved capability III systems. The following lists the major capability upgrades these systems provide.

Improved Capability II

E-36. The improved capability II program was initiated in the 1980s. It was completed across the fleet of EA-6B aircraft (including U.S. Marine Corps aircraft) in the 1990s. The program incorporated incremental capability improvements that include communications, navigation, and computer interface upgrades; a high-speed antiradiation missile capability; and improved jamming pods. Several system interfaces were also upgraded in preparation for the improved capability III improvements.

Improved Capability III

E-37. The improved capability III program incorporates a highly evolved receiver system and provides upgraded EA-6B aircraft with increased signal detection, geolocation capability, a new selective reactive-jamming capability, and better reliability. High-speed antiradiation missile employment is also improved due to the speed of the receiver and its geolocation accuracy. Increased battlefield situational awareness of joint forces is also provided through Link-16. The improved capability III program provides a new ALQ-218 receiver system, integration of the USQ-113 and the multifunctional information distribution system (often called MIDS). This system incorporates Link-16 and various connectivity avionics into the Prowler. The major EW-related subsystems are the AN/ALQ-99 (V) tactical jamming countermeasures set and AN/USQ-113 (V) radio countermeasures set.

E-38. The AN/ALQ-99 (V) tactical jamming countermeasures set has upgraded receivers and processors to provide the following:

- Improved frequency coverage.
- Direction-of-arrival determination capability.
- Narrower frequency discrimination to support narrowband jamming.
- Enhanced interface with onboard systems.

E-39. The AN/USQ-113 (V) radio countermeasures set will enhance the aircraft's jamming capability through its integration with the tactical display system. This will enable the crew to display AN/USQ-113 communications jamming data as well as control AN/USQ-113 operations through the tactical display system.

E/A-18G Growler

E-40. The E/A-18G Growler is the Navy's replacement aircraft for the EA-6B Prowler. Paragraphs E-41 and E-42 discuss the missions, primary tasks, and capabilities of the Navy's E/A-18G Growler. (See figure E-9.) E/A-18G fielding began in 2008. The first operational E/A-18G deployment will occur in 2009, as the Navy begins to replace its carrierborne EA-6B aircraft.

Figure E-9. EA-18 Growler

Mission and Tasks

E-41. The EA-18G can detect, identify, locate, and suppress hostile emitters. It will provide enhanced connectivity to national, combatant command, and strike assets. Additionally, the EA-18G will provide organic accurate emitter targeting using on-board suppression weapons, such as the high-speed antiradiation missile.

Capabilities

E-42. The following is a list of the E/A-18G's general capabilities:

- Suppression of enemy air defenses. The EA-18G will counter enemy air defenses using both reactive and preemptive jamming techniques.
- Stand-off and escort jamming. The EA-18G will be highly effective in the traditional stand-off jamming mission, but with the speed and agility of a Super Hornet, it will also be effective in the escort role.
- Integrated air and ground airborne electronic attack. Enhanced situational awareness and uninterrupted communications will enable the EA-18G to achieve a higher degree of integration with ground operations than previously.
- Self-protect and time-critical strike support. With its active electronically scanned array radar, digital data links, and air-to-air missiles, the EA-18G will be able to protect itself and effectively identify and prosecute targets.
- Growth. High commonality with the F/A-18E and F/A-18F, nine available weapon stations, and modern avionics enable cost-effective synergistic growth, setting the stage for continuous capability enhancement.

E-43. The following is a list of the E/A-18G's airborne electronic attack capabilities:

- Entire spectrum. The EA-18G's ALQ-218 wideband receiver combined with the ALQ-99 tactical jamming system will be effective against any surface-to-air threat.
- Precision airborne electronic attack. Selective-reactive technology enables the EA-18G to rapidly sense and locate threats much more accurately than before. This improved accuracy enables greater concentration of energy against threats.
- Advanced communication countermeasures. Its modular communication countermeasure set enables the EA-18G to counter a wide range of communication systems and is readily adaptable to an ever changing threat spectrum.
- Interference cancellation system. This system dramatically enhances aircrew situational awareness by enabling uninterrupted communications during jamming operations.

CAPABILITIES SUMMARY

E-44. Table E-1 lists Army and joint EW capabilities. (Bold text indicates capabilities not described in the preceding paragraphs.) EW officers, noncommissioned officers, and supporting staff members should be familiar with these capabilities and how they can support Army operations. Additional information on the EW capabilities listed in table E-1 is found in the Web sites listed in table E-2, page E-12.

Table E-1. Army and joint electronic warfare capabilities

	Army	Air Force	Navy	Marine Corps
Airborne	RC-12 Guardrail	**EC-130J Commando Solo**	EA-6B Prowler	EA-6B Prowler
	airborne common sensor	EC-130H Compass Call	EA-18G Growler	
		RC-135V/W Rivet Joint	**EP-3E Aries II**	
		F-16CJ		
		E-8 JSTARS		
Unmanned aircraft system*	**RQ-5A/MQ-5B Hunter (Corps)**	**RQ-4A (Joint) Global Hawk**	**RQ-2 Pioneer**	
	RQ-7A/B Shadow (brigade)	**RQ-1L (Joint) Predator**	**MQ-8B Fire Scout Vertical Take-off**	
	MQ-1C/Sky Warrior (replacement for Hunter)	**RQ-11 Raven**	**Silver Fox**	
	MQ-8 Fire Scout			**RQ-11 Raven**
	RQ-11 Raven (battalion) Hand Launched			**Scan Eagle**
Ground	AN/MLQ-40 Prophet		**CREW Systems (Joint)**	AN/MLQ-36 MEWSS

Note: *Other Services may refer to unmanned aircraft systems as unmanned aerial systems or vehicles.
CREW counter radio-controlled improvised explosive device electronic warfare
MEWSS mobile electronic warfare support system
SOF special operations forces

Table E-2. Electronic warfare systems and platforms resources

Army platforms and systems http://www.sed.monmouth.army.mil/avionics/ http://www.sec.army.mil/secweb/fact_sheets/fact_sheets.php
Air Force platforms and systems http://www.af.mil/factsheets/factsheet.asp?fsID=182 http://www.airforce-technology.com/projects/#Unmanned_Aerial_Vehicles_(UAV_/_UCAV)
Navy systems platforms and systems http://acquisition.navy.mil/programs http://www.naval-technology.com/projects/ http://www.navy.mil/navydata/fact.asp
Marine Corps platforms and systems http://www.marcorsyscom.usmc.mil/sites/cins/INTEL/USMC%20CREW/index.html
Joint programs https://www.jieddo.dod.mil

Appendix F

Tools and Resources Related to Electronic Warfare

This appendix provides information on tools and reachback resources related to electronic warfare. Electronic warfare officers, noncommissioned officers, and supporting staff members should be familiar with these tools and resources and how to use them to support electronic warfare operations. Some tools and resources require an approved user account prior to being granted access.

ARMY REPROGRAMMING ANALYSIS TEAM

F-1. The Army Reprogramming Analysis Team (ARAT) supports tactical commanders. It provides timely reprogramming of any Army-supported software used for target acquisition, target engagement, measurement and signature intelligence, and vehicle and aircraft survivability (including that operated by other Services). The team provides software changes not readily possible by operator input to respond to rapid deployments or changes in the operational environment. See their Web site at https://ako.sec.army.mil/arat/index.html (Army Knowledge Online login required).

F-2. ARAT provides reprogramming support to counter-radio-controlled improvised-explosive-device (IED) electronic warfare (EW) (sometimes referred to as CREW), and other electronic systems.

F-3. The team is accessible via the Army Reprogramming Analysis Team's Warfighter Survivability Software Support Portal. A secure Internet protocol router network (SIPRNET) account is required to access the portal.

NATIONAL GROUND INTELLIGENCE CENTER

F-4. The National Ground Intelligence Center provides all-source analysis of the threat posed by IEDs produced and used by foreign terrorist and insurgent groups. The center supports U.S. forces during training, operational planning, deployment, and redeployment.

F-5. The center maintains a counter-IED targeting program (often called CITP) portal on its SIPRNET site. This portal provides information concerning IED activities and incidents as well as IED assessments.

ELECTRONIC ORDER OF BATTLE

F-6. An electronic order of battle details all known combinations of emitters and platforms in a particular area of responsibility. It consists of several reachback resources:

- National Security Agency-Electronic Intelligence Parameter Query.
- U.S. electromagnetic systems database.
- National Ground Intelligence System parametric information relational intelligence tool database.
- Military equipment parametrics and engineering database.

E-SPACE

F-7. E-Space is a Department of Defense (DOD) entity housed in the National Security Agency. It provides intelligence assistance (primarily signals intelligence) to deployed EW officers. E-Space is a reachback capability available to EW officers and spectrum managers that can be leveraged to provide all-source intelligence products and answers to requests for information and spectrum interference questions.

JOINT ELECTRONIC WARFARE CENTER

F-8. The Joint Electronic Warfare Center is DOD's only joint EW center of expertise. It provides EW subject matter expertise from a range of backgrounds, including people with current multi-Service operational experience. The center has a limited capability to perform modeling and simulation studies and EW red team support. It can deploy in a support role if approved by the U.S. Strategic Command.

JOINT IMPROVISED EXPLOSIVE DEVICE DEFEAT ORGANIZATION

F-9. The Joint Improvised Explosive Device Defeat Organization (known as JIEDDO) leads, advocates, and coordinates all DOD actions in support of efforts by combatant commanders and their joint task forces to defeat IEDs as weapon of strategic influence.

JOINT SPECTRUM CENTER

F-10. The Joint Spectrum Center ensures DOD effectively uses the electromagnetic spectrum in support of national security and military objectives. The center serves as DOD's center of excellence for electromagnetic spectrum management matters in support of the combatant commands, military departments, and DOD agencies in planning, acquisition, training, and operations.

F-11. The center maintains databases and provides data about friendly force command and control information system locational and technical characteristics. This information is used to plan electronic protection measures. These databases provide EW planners with information covering communication, radar, navigation, broadcast, identification, and EW systems operated by the DOD, other government agencies, and private businesses and organizations.

F-12. The center provides information on a quick-reaction basis in various formats and media to support EW planners and spectrum managers.

KNOWLEDGE AND INFORMATION FUSION EXCHANGE

F-13. The Knowledge and Information Fusion Exchange (sometimes called KnIFE) is a program sponsored by U.S. Joint Forces Command. It provides Soldiers with observations, insights, and lessons from operations around the world.

ADDITIONAL INFORMATION

F-14. Further information on the above tools and resources can be accessed through Army Knowledge Online. The links to these Web sites can be viewed by first accessing the "Army Operational Electronic Warfare Course" on Army Knowledge Online at http://www.us.army.mil/suite/page/400055 and then clicking on Folders >Links>EW links.

Glossary

SECTION I – ACRONYMS AND ABBREVIATIONS

ARAT	Army Reprogramming Analysis Team
C-3	operations directorate of a multinational (combined) staff
CJCSI	Chairman of the Joint Chiefs of Staff instruction
CJCSM	Chairman of the Joint Chiefs of Staff manual
COA	course of action
DD	Department of Defense (official forms only)
DOD	Department of Defense
DODI	Department of Defense Instruction
EW	electronic warfare
FM	field manual
FMI	field manual, interim
G-2	assistant chief of staff, intelligence
G-3	assistant chief of staff, operations
G-5	assistant chief of staff, plans
G-6	assistant chief of staff, signal
G-7	assistant chief of staff, information engagement
GTA	graphic training aid
HF	high frequency
Hz	hertz
IED	improvised explosive device
IO	information operations
IPB	intelligence preparation of the battlefield
ISR	intelligence, surveillance, and reconnaissance
J-2	intelligence directorate of a joint staff
J-3	operations directorate of a joint staff
J-5	plans directorate of a joint staff
J-6	communications system directorate of a joint staff
JFMO	Joint Frequency Management Office
JIOWC	Joint Information Operations Warfare Center
JP	joint publication
MAGTF	Marine air-ground task force
MC	Military Committee (NATO)
MCWP	Marine Corps warfighting publication
MDMP	military decisionmaking process

NATO	North Atlantic Treaty Organization
S-2	intelligence staff officer
S-3	operations staff officer
S-6	signal staff officer
S-7	information engagement staff officer
SIPRNET	SECRET Internet Protocol Router Network
STANAG	standardization agreement (NATO)
TERPES	tactical electronic reconnaissance processing and evaluation system
U.S.	United States

SECTION II – TERMS

communications security

(joint) The protection resulting from all measures designed to deny unauthorized persons information of value that might be derived from the possession and study of telecommunications, or to mislead unauthorized persons in their interpretation of the results of such possession and study. (JP 6-0)

computer network operations

(joint) Comprised of computer network attack, computer network defense, and related computer network exploitation enabling operations. (JP 3-13)

directed energy

(joint) An umbrella term covering technologies that relate to the production of a beam of concentrated electromagnetic energy or atomic or subatomic particles. (JP 3-13.1)

electromagnetic environment

(joint) The resulting product of the power and time distribution, in various frequency ranges, of the radiated or conducted electromagnetic emission levels that may be encountered by a military force, system, or platform when performing its assigned mission in its intended operational environment. It is the sum of the electromagnetic interference; electromagnetic pulse; hazards of electromagnetic radiation to personnel, ordnance, and volatile materials; and natural phenomena effects of lightning and precipitation static. (JP 3-13.1)

electromagnetic environmental effects

The impact of the electromagnetic environment upon the operational capability of military forces, equipment, systems, and platforms. It encompasses all electromagnetic disciplines, including electromagnetic compatibility and electromagnetic interference; electromagnetic vulnerability; electromagnetic pulse; electronic protection, hazards of electromagnetic radiation to personnel, ordnance, and volatile materials; and natural phenomena effects of lightning and precipitation static. (JP 3-13.1)

electromagnetic spectrum

(joint) The range of frequencies of electromagnetic radiation from zero to infinity. It is divided into 26 alphabetically designated bands. (JP 1-02)

electromagnetic vulnerability

(joint) The characteristics of a system that cause it to suffer a definite degradation (incapability to perform the designated mission) as a result of having been subjected to a certain level of electromagnetic environmental effects. (JP 1-02)

electronic attack

(joint) Division of electronic warfare involving the use of electromagnetic energy, directed energy, or antiradiation weapons to attack personnel, facilities, or equipment with the intent of degrading, neutralizing, or destroying enemy combat capability and is considered a form of fires. (JP 3-13.1)

electronic protection

(joint) Division of electronic warfare involving actions taken to protect personnel, facilities, and equipment from any effects of friendly or enemy use of the electromagnetic spectrum that degrade, neutralize or destroy friendly combat capability. (JP 3-13.1)

electronic warfare

(joint) Military action involving the use of electromagnetic and directed energy to control the electromagnetic spectrum or to attack the enemy. Electronic warfare consists of three divisions: electronic attack, electronic protection, and electronic warfare support. (JP 3-13.1)

electronic warfare support

(joint) Division of electronic warfare involving actions tasked by, or under direct control of, an operational commander to search for, intercept, identify, and locate or localize sources of intentional and unintentional radiated electromagnetic energy for the purpose of immediate threat recognition, targeting, planning, and conduct of future operations. (JP 3-13.1)

emission control

(joint) The selective and controlled use of electromagnetic, acoustic, or other emitters to optimize command and control capabilities while minimizing, for operations security: a. detection by enemy sensors; b. mutual interference among friendly systems; and/or c. enemy interference with the ability to execute a military deception plan. (JP 1-02)

joint restricted frequency list

(joint) A time a geographically-oriented listing of TABOO, PROTECTED, and GUARDED functions, nets, and frequencies. It should be limited to the minimum number of frequencies necessary for friendly forces to accomplish objectives. (JP 3-13.1)

working group

(Army) A temporary grouping of predetermined staff representatives who meet to coordinate and provide recommendations for a particular purpose or function. (FMI 5-0.1)

References

REQUIRED PUBLICATIONS

These documents must be available to intended users of this publication.

FM 1-02 (101-5-1). *Operational Terms and Graphics*. 21 September 2004.

JP 1-02. *Department of Defense Dictionary of Military and Associated Terms*. 12 April 2001. (As amended through 4 March 2008.)

JP 3-13.1. *Electronic Warfare*. 25 January 2007.

FM 3-0. *Operations*. 27 February 2008.

FM 5-0 (101-5). *Army Planning and Orders Production*. 20 January 2005.

FM 6-0. *Mission Command: Command and Control of Army Forces*. 11 August 2003.

FMI 5-0.1. *The Operations Process*. 31 March 2006.

RELATED PUBLICATIONS

These documents contain relevant supplemental information.

JOINT AND DEPARTMENT OF DEFENSE PUBLICATIONS

Most joint publications are available online: <http://www.dtic.mil/doctrine/jpcapstonepubs.htm.>

CJCSI 3320.01B *Electromagnetic Spectrum Use in Joint Military Operations*. 01 May 2005

CJCSI 3320.02C. Joint Spectrum Interference Resolution (JSIR). 27 January 2006 (with change 1 as of 25 February 2008).

CJCSI 3320.03A *Joint Communications Electronics Operation Instructions*. 11 June 2005.

CJCSM 3122.03C. *Joint Operation Planning and Execution System Volume II, Planning Formats and Guidance*. 17 August 2007.

CJCSM 3320.01B *Joint Operations in the Electromagnetic Battlespace*. 25 March 2006.

CJCSM 3320.02A *Joint Spectrum Interference Resolution (JSIR) Procedures*. 16 February 2006.

DODI 4650.01. Policy and Procedures for Management and Use of the Electromagnetic Spectrum. 09 January 2009.

JP 2-0. *Joint Intelligence*. 22 June 2007.

JP 2-01. *Joint and National Intelligence Support to Military Operations*. 07 October 2004.

JP 3-0. *Joint Operations*. 17 September 2006.

JP 3-09. *Joint Fire Support*. 13 November 2006.

JP 3-09.3. *Joint Tactics, Techniques, and Procedures for Close Air Support (CAS)*. 03 September 2003.

JP 3-13. *Information Operations*. 13 February 2006.

JP 3-13.3. *Operations Security*. 29 June 2006.

JP 3-13.4 (JP 3-58). *Military Deception*. 13 July 2006.

JP 3-60. *Joint Targeting*. 13 April 2007.

JP 6-0. *Joint Communications System*. 20 March 2006.

ARMY PUBLICATIONS

Most Army doctrinal publications are available online: <http://www.army.mil/usapa/doctrine/Active_FM.html>.

FM 2-0 (34-1). *Intelligence.* 17 May 2004.

FM 3-09.32. JFIRE: *Multi-Service Tactics, Techniques, and Procedures for the Joint Application of Firepower*. 20 December 2007.

FM 3-13 (100-6). *Information Operations: Doctrine, Tactics, Techniques, and Procedures.* 28 November 2003.

FM 3-13.10 (3-51.1). *Multi-Service Tactics, Techniques, and Procedures for the Reprogramming of Electronic Warfare and Target Sensing Systems*. 22 January 2007.

FM 5-19 (100-14). *Composite Risk Management*. 21 August 2006.

FM 6-20-10. *Tactics, Techniques, and Procedures for the Targeting Process.* 8 May 1996.

FM 6-99.2 (101-5-2). *U.S. Army Report and Message Formats*. 30 April 2007.

FM 34-130. *Intelligence Preparation of the Battlefield.* 8 July 1994.

FMI 2-01. *Intelligence, Surveillance, and Reconnaissance (ISR) Synchronization.* 11 November 2008.

GTA 90-10-046. *MNC-I Counter IED Smart Book.* September 2008.

NATO PUBLICATIONS

Allied Joint Publication 3.6. *Allied Joint Electronic Warfare Doctrine*. December 2003.

MC 64. *NATO Electronic Warfare (EW) Policy*. 26 April 2004.

STANAG 5048 C3 (Edition 5). *The Minimum Scale of Convectivity for Communications and Information Systems for NATO Land Forces*. 16 February 2000.

OTHER PUBLICATIONS

AFDD 2-1.9. *Targeting*. 8 June 2006.

AFDD 2-5.1. *Electronic Warfare*. 5 November 2002.

Executive Order 12333. *United States Intelligence Activities*. 4 December 1981.

MCWP 2-22 (2-15.2). *Signals Intelligence*. 13 July 2004.

NWP 3-13. *Navy Information Operations*. June 2003.

SOURCES USED

Electronic Warfare Working Group, U.S. House of Representatives, Issue Brief #17. "Compass Call During Operation Iraqi Freedom." 11 March 2004. Available online at http://www.house.gov/pitts/initiatives/ew/Library/Briefs/brief17.htm.

PRESCRIBED FORMS

None

REFERENCED FORMS

DA Forms are available on the APD website (www.apd.army.mil. DD forms are available on the OSD website (www.dtic.mil/whs/directives/infomgt/forms/formsprogram.htm).

DA Form 2028. *Recommended Changes to Publications and Blank Forms.*

DD Form 1972. *Joint Tactical Air Strike Request.*

Index

Entries are by paragraph number unless specified otherwise.

Entries are by paragraph number unless specified otherwise.

Entries are by paragraph number unless specified otherwise.

Entries are by paragraph number unless specified otherwise.

Entries are by paragraph number unless specified otherwise.

Entries are by paragraph number unless specified otherwise.